Education 2014

Core Programming Area at the 2014 AIChE Spring Meeting and 10th Global Congress on Process Safety

New Orleans, Louisiana, USA
30 March – 3 April 2014

ISBN: 978-1-63439-069-9

Printed from e-media with permission by:

Curran Associates, Inc.
57 Morehouse Lane
Red Hook, NY 12571

CURRAN ASSOCIATES INC.
proceedings.com

Some format issues inherent in the e-media version may also appear in this print version.

Printed by Curran Associates, Inc. (2014)

For permission requests, please contact AIChE
at the address below.

AIChE
3 Park Avenue
New York, NY 10016-5991

Phone: (203) 702-7660
Fax: (203) 775-5177

www.aiche.org

Additional copies of this publication are available from:

Curran Associates, Inc.
57 Morehouse Lane
Red Hook, NY 12571 USA
Phone: 845-758-0400
Fax: 845-758-2634
Email: curran@proceedings.com
Web: www.proceedings.com

TABLE OF CONTENTS

354012 Possible vs. Practical: Engineers Must Lead the Development of Practical Technologies

Abstract:

Chemical Engineering is need today more than ever. There is a strong societal desire for sources of energy and feedstocks that can sustainably support a high standard of living for all. Thermodynamic, material and economic challenges stand in the way. Most of the general public does not have the basic understanding of the energy flows, chemical transformations, and the scale of the systems that support our current way of life. These concepts are fundamental to chemical engineering. Chemical engineers should be on the vanguard of separating hype from truly practical options yet occationally we let society down. This talk will present several examples where the fundamentals of chemical engineering can be used to access the practicality of proposed solutions to energy and water issues, as well as explain why the venture capital model struggles in the materials market.

353750 Introduction to Project Management Fundamentals

Abstract:

This presentation introduces the fundamentals of project management. It will be especially useful for new engineers, young professionals, and advanced students, as well as anyone who serves or will served on project teams or leads those teams. We will discuss the importance of people, the importance of setting excellent project objectives, and the importance of good planning. Working with people is perhaps the most difficult. Suggestions will be given on how to better communicate and better achieve unity with project team members. We will discuss the real meaning of the word "Communication" and how that meaning can change the whole approach we take to interacting with people in a productive way. Important ideas and tools for developing excellent project objectives will be presented, including the use of SMART objectives, the triple constraints, and tolerance diagrams. An outline of the project planning process will be reviewed. This planning process includes the use of work breakdown structures, network diagrams, and Gantt charts, as well as resource histograms. Also, we will briefly discuss doing fatal-flaw analysis.

353751 Becoming a Real Team to Accomplish Projects

Abstract:

In project management a major effort goes into becoming a real team! A team is not just a group of individuals! It is a very special group of individuals who have worked and learned to become united in their objectives and willingness to work to accomplish those objectives. This paper will discuss in depth the important principles of people interaction skills needed for people to coalesce into a working unit of synergy. We will discuss communication, what it means, and how to use it. We will talk about leading effective team meetings. We will discuss how to look at issues from other people's perspectives in order to better understand them and work with them to accomplish objectives. We will discuss listening and barriers to effective listening. All of this will go toward developing a new vision of what it means to be an effective and united team and give important meaning to the word "teamwork."

353907 Project Management:Things You Should Think About

Abstract:

"Beyond the critical fundamentals of communication, working with people, setting good objectives, and the importance of good planning, there are some other topics that are important for project managers to consider. Some of these topics are managing risks, managing conflicts, dealing with contracts, managing meetings, situational leadership, taking control of yourself, and so on. This presentation will briefly talk about many of these aspects—each one is really a major topic by itself, but we will give some key helpful hints about some things you should consider as you begin working on project teams, become a new project manager, or perhaps brush up on things to consider even as an experienced project manager. This paper will be helpful to all of these people. We'll give some important rules of effective project management, some key ideas about managing conflict, and some basics on managing risk, and hopefully aid you in your path toward effective project management!"

348790 Leadership Starts Now!

Abstract:

Often people think that leadership begins when there is a promotion to a position of authority with direct personnel responsibilities. Some believe that leadership can simply be switched on after assuming a title. Nothing could be further from the truth and to follow that prescription can be detrimental for young professionals. As new engineers come into an organization, it is essential they understand the informal opportunities and subtle behaviors that reveal their leadership aptitude.

This presentation will discuss the "hidden leadership" opportunities that are woven into your everyday roles, how to recognize these opportunities and take advantage of them. Don't wait until you are in a formal leadership position. Learn how to be a leader at every opportunity; starting now!

The main target audience is entry level engineers and professionals who are beginning their careers across all industry types.

During this presentation you will learn:

- The difference between managing and leading
- Qualities and behaviors that are recognized as leadership
- Common errors that result in missed opportunities for leadership
- Methods for leading from a non-leadership role or position
- How to recognize "hidden leadership" opportunities
  Steps you should take NOW to develop yourself as a leader

353909 Panel Discussion-What Do You Want to Know About Project Management

Abstract:

: "The session participants will have an opportunity to ask questions of experienced project managers on how to be better project managers and any other questions they have respecting the presentations of this session."

* DESMAR SOLAR

NATALY AGUDELO
YURANI GUTIÉRREZ
ANDRES OLAYA
FELIPE SASTOQUE

Energetic and exergetic analisys from a seawater solar desalinator

DESMAR SOLAR

*SCHEDULE

- Introduction
- Justification
- Objectives
- Resources available at the university.
- Main processes of water desalination.
- Material and energy balances.
- Preliminary test.
- Environmental considerations.
- Results
- Bibliography.

DESMAR SOLAR

*INTRODUCTION

Water is the main element that humans use in their daily work. Of all the water available on the planet only 3% is fresh water and is on the surface. The remaining water is either frozen at the poles or distributed in the oceans.

DESMAR SOLAR

*JUSTIFICATION

Desalination processes are quite costly and cause pollution to the environment because their main source of energy is at its most fossil fuels.

Therefore, this investigation is intended to use efficient and clean energy sources to develop the process

DESMAR SOLAR

*COLOMBIAN SOLAR MAP

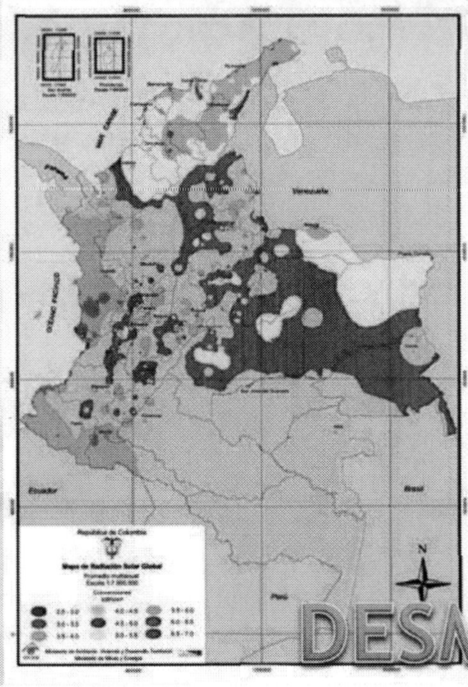

DESMAR SOLAR

* COLOMBIAN MAIN AREAS AFFECTED

San Pedro-Sucre

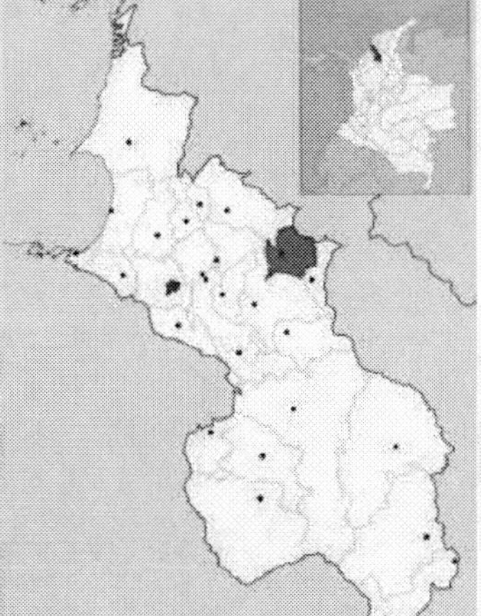

Montecristo-Bolívar

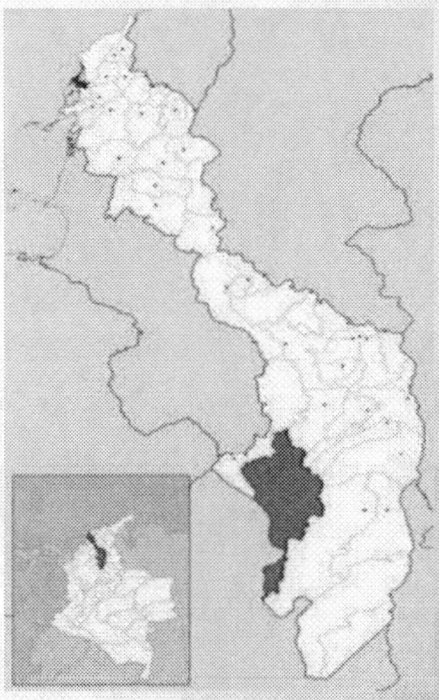

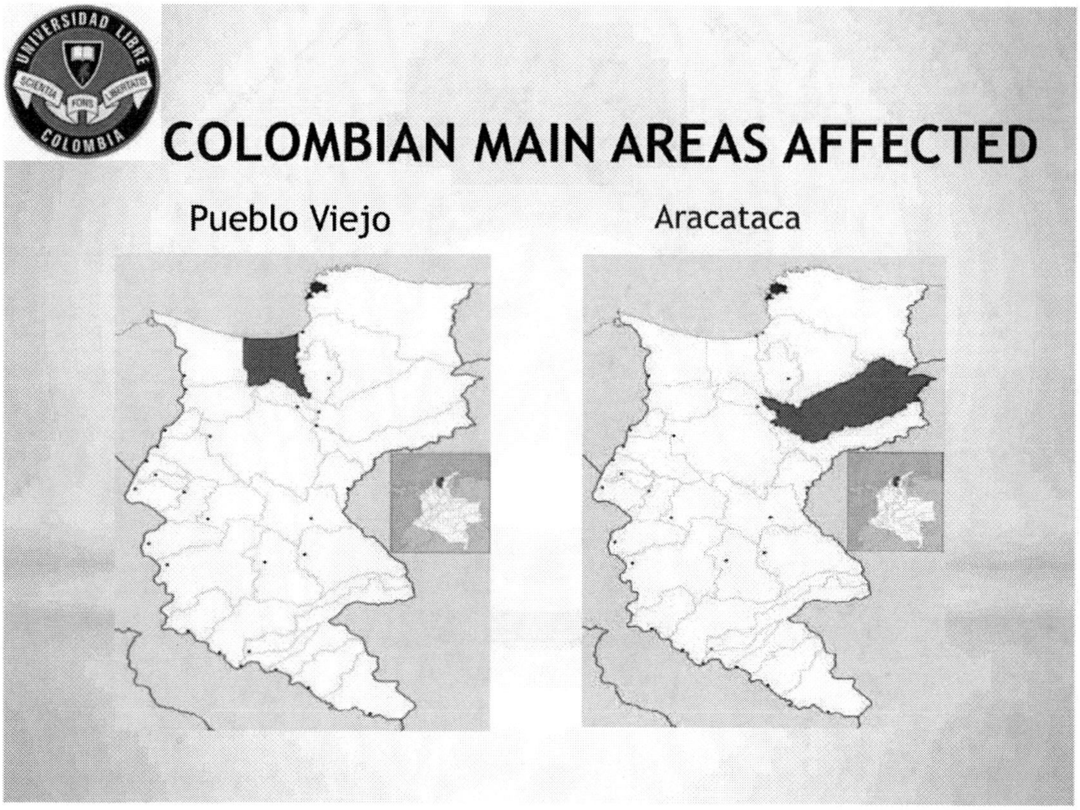
UNIVERSIDAD LIBRE
SCIENTIA FONS LIBERTATIS
COLOMBIA
COLOMBIAN MAIN AREAS AFFECTED
Pueblo Viejo
Aracataca

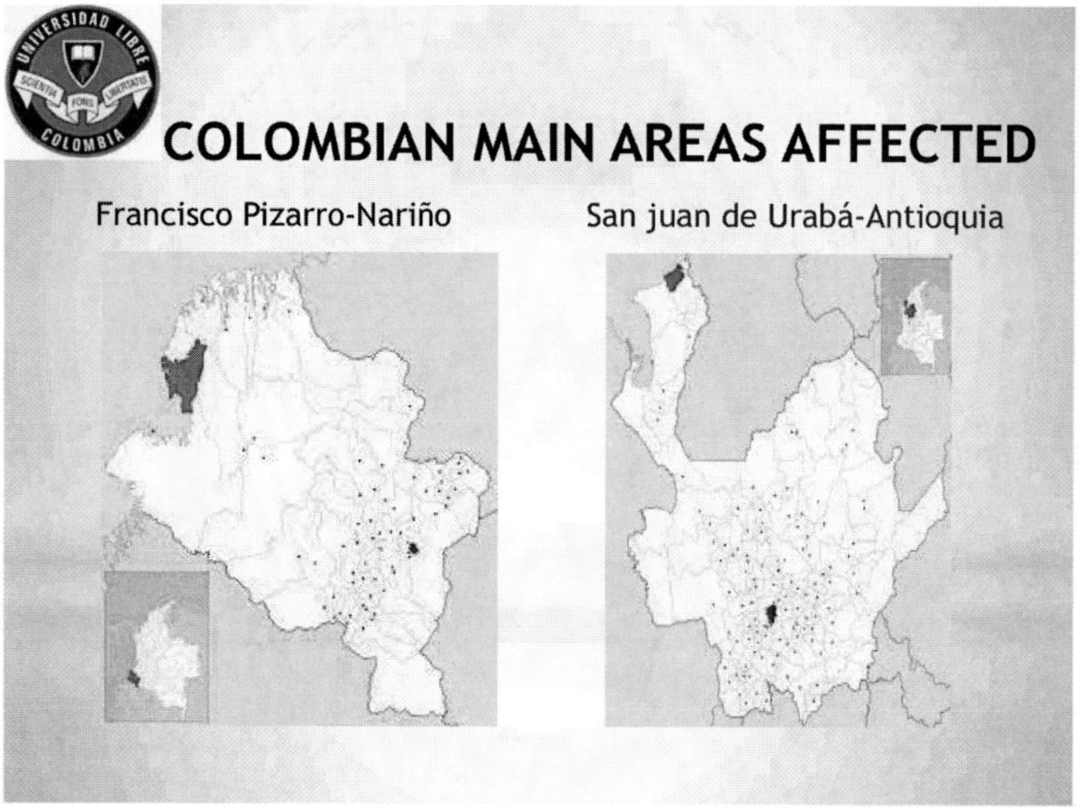
UNIVERSIDAD LIBRE
SCIENTIA FONS LIBERTATIS
COLOMBIA
COLOMBIAN MAIN AREAS AFFECTED
Francisco Pizarro-Nariño
San juan de Urabá-Antioquia

COLOMBIAN MAIN AREAS AFFECTED

Timbiquí

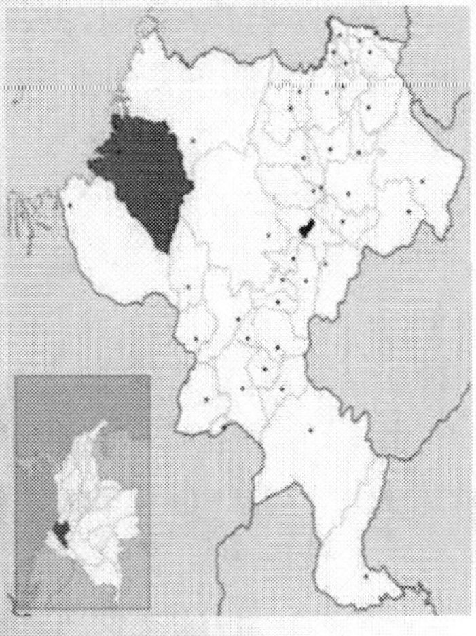

Villa rica

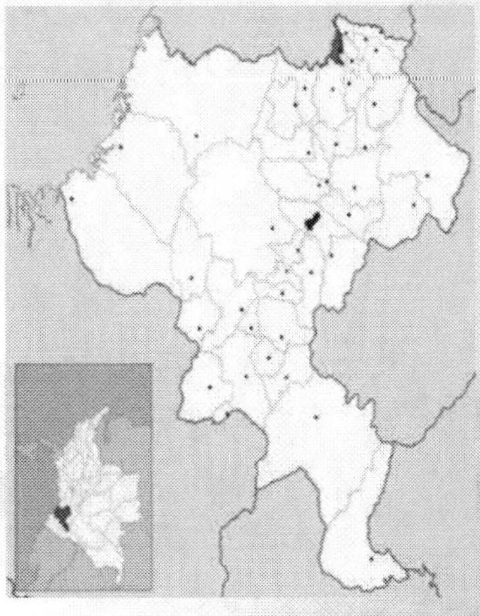

* GENERAL PURPOSE

To establish the energy and exergy analysis of a desalination system using solar energy.

DESMAR SOLAR

*SPECIFIC PURPOSE

*To determine the energy requirements of each of the stages of a solar desalination process.

*To establish the efficiency of different models of solar energy collection system to be used in the desalination process.

*To evaluate the exergy of the process and establish improvements in the design selecting the most suitable model system for the process

*To determine physical properties of the materials.

*To quantify the effect of temperature and the properties of the materials for their selection and use in a solar desalination.

*To establish the lifetime of the material chosen from the tests already performed.

DESALINATION PROCESS

*Desalination by reverse osmosis

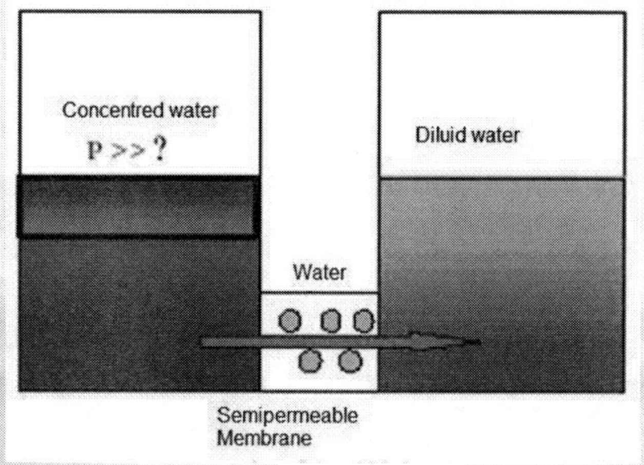

DESMAR SOLAR

*Desalination by Freezing

*Desalination by hydrate formation

*Flash point desalination

DESMAR SOLAR

*Desalination by distillation

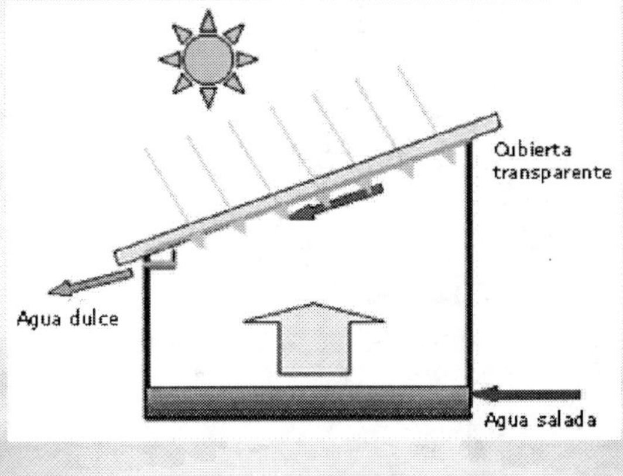

DESMAR SOLAR

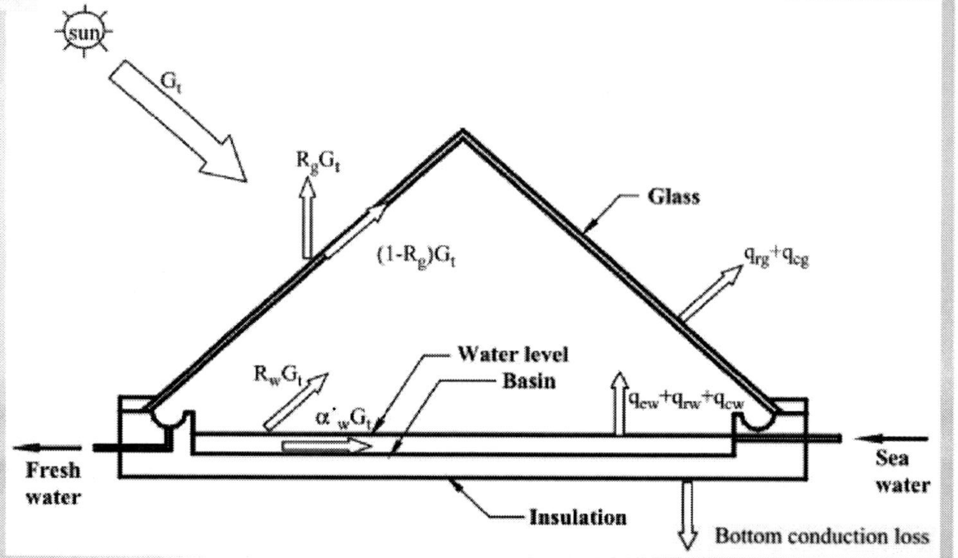

DESMAR SOLAR

*Energetic and exergetic models

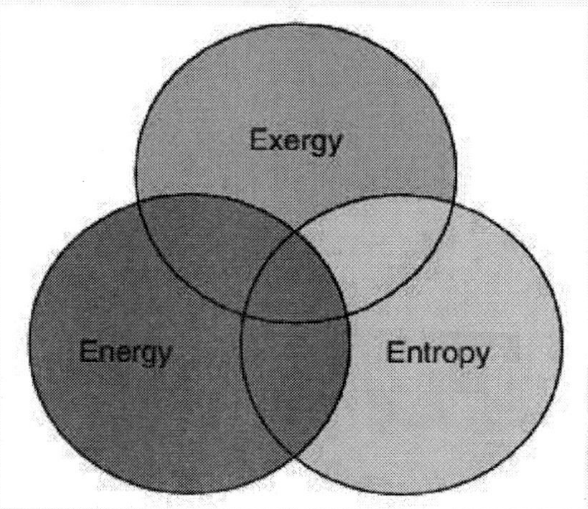

DESMAR SOLAR

ENERGY AND EXERGY DEFINITION?

*Energy: The ability that have the matter to produce work in the form of motion, light, heat, etc..

*Exergy: A thermodynamic property that determines the useful work potential of a given amount of energy that can be achieved by the spontaneous interaction between a system and its environment. Reports the system's potential usefulness as a source of labor.

DESMAR SOLAR

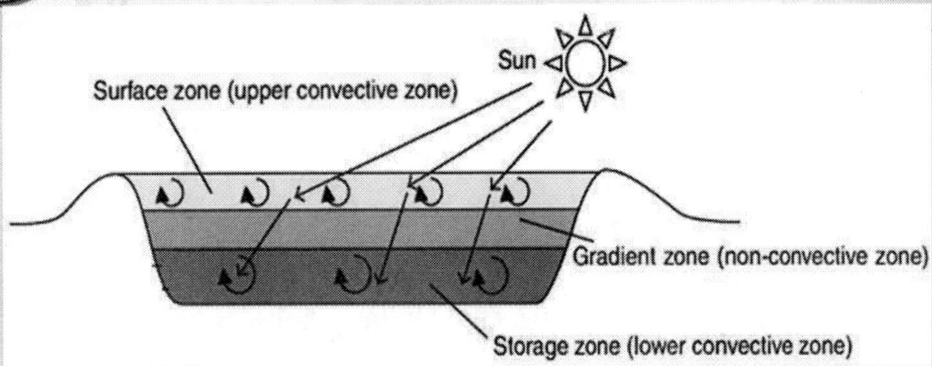

$$n = \frac{m_s h_s + m_w h_w}{G_t}$$

*EXERGY

$$Q_{net} = Q_{HSZ,solar} - Q_{bottom} - Q_{up} - Q_{side}$$

$$n_{HSZ} = 1 - \frac{(Q_{bottom} + Q_{up} + Q_{side})}{Q_{HSZ,solar}}$$

DESMAR SOLAR

*ENTROPY

$$\begin{aligned} s &= x_s s_s + x_w s_w \\ &= x_s[s_{s,\mathrm{pure}}(T,P) - R\ln(x_s)] + x_w[s_{w,\mathrm{pure}}(T,P) - R\ln(x_w)] \\ &= x_x s_{s,\mathrm{pure}}(T,P) - R[x_s \ln(x_s) + x_w \ln(x_w)] \end{aligned}$$

DESMAR SOLAR

*EXPERIMENTS

*EXPERIMENT 1.

Distillation through heating plate

DESMAR SOLAR

* HEATING DATA

	TEMPERATURE (°C)	TIME (min)
Evaporated	67	02:46
Finished	91	43:33

FLOW COOLING

Measure	Volume (mL)	Time(s)	Cooling flow(mL/seg)	Cooling flow(L/min)
1	1000	25,57	39,1	2,35
2	1000	25,86	38,7	2,32
3	1000	25,26	39,6	2,38
	PROMEDIO	25,56	39,1	2,35

DESMAR SOLAR

RESULTS

Solution	Temperature (°C)	Volume (mL)	ppm
Initial	20	200	$90x10^3$
Final	20	128	$148x10^3$
Desalinated	20,4	68	0

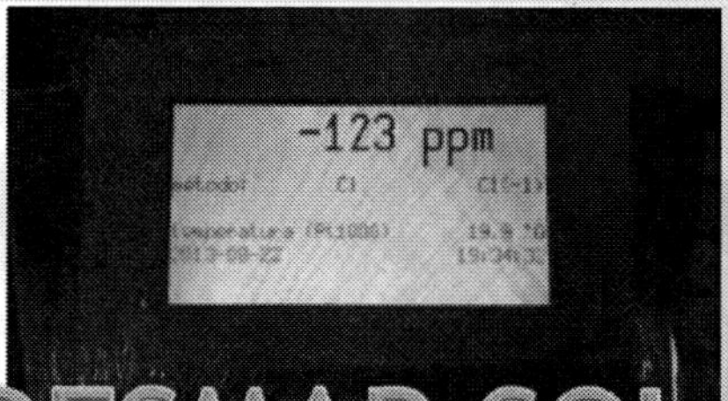

DESMAR SOLAR

BALANCE RESULTS

1

Matter balance		
Input	Output	
200 ml	128 mL	68 mL
90×10^3	148×10^3	0

me = m1+ m2	$90x10^{3}$ * 0,200= 0 * 0,068 + 0,128 *ppm output
ppm output=	140×10^{3}

PERCENTAGE OF ERROR

$$\frac{ppm_{plancha} ppm_{\ salida}}{ppm_{plancha}} * 100 \qquad \frac{148000 - 140000}{148000} * 100 = 5.40\%$$

DESMAR SOLAR

2

Energy Balance		
	w*t	
Input	2,8 x 10^7	W * seg
	7,8	kWh

$$\rho v C_{H2O} \Delta T = m \Delta H + m C_{H2O} \Delta T$$

ρ	1037,5	kg/m^3
v	0,0002	m^3
C H2O	4184	kJ/(seg * K)
ΔT	70,68	K
m	0,2075	kg
ΔH	173	kJ/(kg * K)
61362	=	61398
	36 J	

DESMAR SOLAR

EXPERIMENT 2

Distillation through solar simulator (with incandescent bulbs)

DESMAR SOLAR

HEATING DATA SAMPLE

	TEMPERATURE (°C)	TIME(min)
Evaporated	38	46
Finished	90	380

DESMAR SOLAR

BALANCE RESULT

Matter balance		
Input	Output	
200 ml	148mL	52 mL
90×10^3	115×10^3	0

me = m1+ m2	$90 \times 10^3 * 0{,}200 = 0 * 0{,}052 + 0{,}148$ *ppm output
ppm output=	$121{,}62 \times 10^3$

PERCENTAGE OF ERROR

$$\frac{ppm_{plancha} ppm_{salida}}{ppm_{plancha}} * 100 \qquad \left|\frac{115000 - 121620}{115000}\right| * 100 = 5{,}76\%$$

DESMAR SOLAR

RESULTS

Solution	Temperature (°C)	Volume (mL)	ppm
Initial	20	200	$90x10^3$
Final	20	148	$121,62x10^3$
Desalinated	54	52	0

DESMAR SOLAR

*MATERIALS FOR THE DESALINATOR

DESMAR SOLAR

*TYPES OF MATERIALS

*Metallic Materials

*Ceramic Materials

*Plastics Materials

DESMAR SOLAR

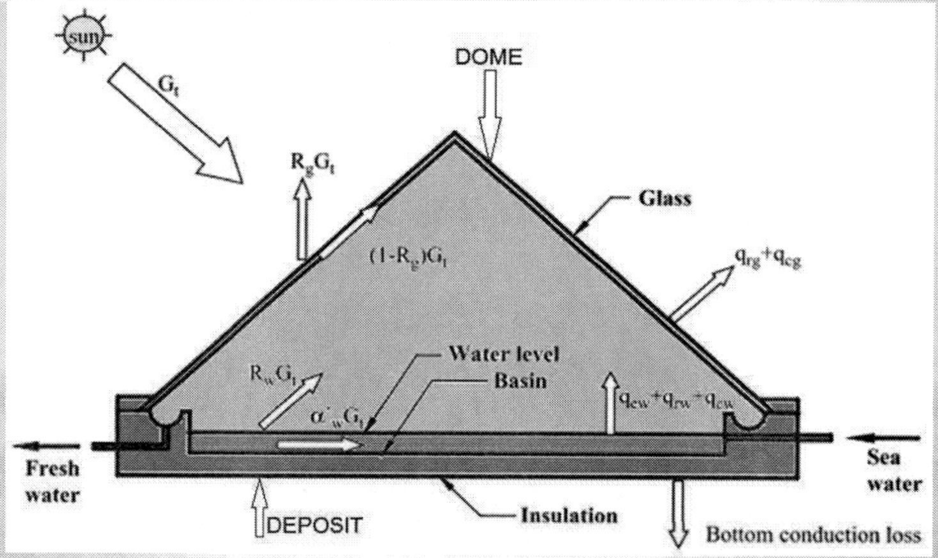

DESMAR SOLAR

*MATERIALS FOR THE DOME

DESMAR SOLAR

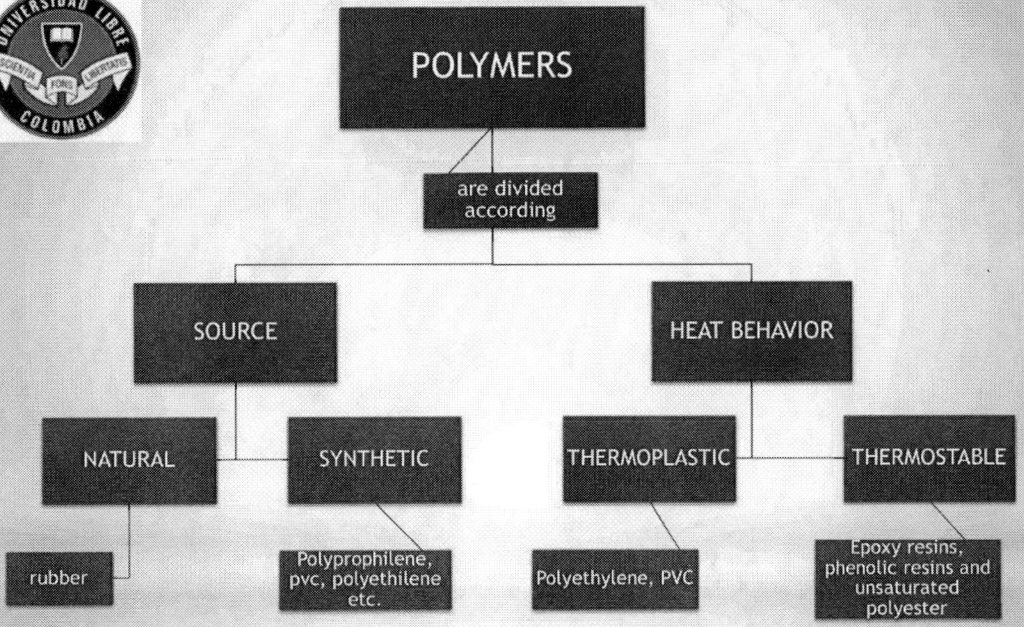

DESMAR SOLAR

*PROPERTIES

*Elongation

*Thermal conductivity

*Specific heat

*Coefficient of thermal expansion

DESMAR SOLAR

	•Polietileno de baja densidad (LDPE)	•Polietileno de alta densidad (HDPE)	•Polipropileno (PP)	•Cloruro de polivinilo rigido(pvc)	•Cloruro de polivinilo plastico(pvc)
Elongation [%]	90-800	50-1000	200-700	002-40	200-450
Thermal conductivity [kw/mk]	3,35	4,6-5,2	1,17	1,3-2,9	1,3-1,7
Specific heat [kj/kgk]	2,3	2,3	1,9	0,84-1,17	1,26-2,09
Coefficient of linear thermal expansion [K^-1]	1e-4 - 2e-4	1,1e-4 - 1,3e-4	5,8-10,2	5-18,5	7-25
Continuous use temperature [K]	355,37-373,15	394,26	394,26-433,15	338,71-352,59	338,71-352,59
Deflection temperature [K]	100-121	140-190	200-230	135-180	…………
Solvents	N/a	N/a	N/a	Cetonas, ésteres,aromaticos,hidrocarburos	Cetonas, ésteres,aromaticos,hidrocarburos
Elasticity in flexure limit [pa]	…………	6894,76	41368,56-55158,08	68947,6-110316,16	N/a

*APPROPRIATE MATERIALS

*Low density polyethylene (LDPE)

*High density polyethylene (HDPE)

*Polypropylene (PP)

*Polyvinyl chloride (PVC)

DESMAR SOLAR

DEPOSIT MATERIALS

DESMAR SOLAR

CERAMIC MATERIALS

* Nonmetallic inorganic solids produced by heat treatment.
* Compared with metals and plastics are hard, non-combustible and non-rusting.
* They can be used in environments with high temperature tribological and corrosive.

DESMAR SOLAR

CERAMIC´S PROPERTIES

* High melting temperature.
* Low density.
* Strength, hardness, structural strength.
* Corrosion resistance.
* Good thermal and electrical insulators.

DESMAR SOLAR

ENVIRONMENTAL CONSIDERATIONS

- By using solar energy production CO2 is removed.
- The use and advancement of renewable energy is implemented. (In our case solar energy is used).
- Drinkable water in coastal areas.

DESMAR SOLAR

*BIBLIOGRAPHY

- Robert Cook (1986). Computer Simulation of Stress-Strain Behavior in Polymeric Materials. MRS Proceedings, 79, 409 doi:10.1557/PROC-79-409.
- David L. Edwards and Mircea Chipara (2004). Polymeric materials for Solar Sail: The combined effects of polymer thickness, radiation, and temperature. MRS Proceedings, 851, NN1.6 doi:10.1557/PROC-851-NN1.6.
- Alice B. Paterakis (1992). The Consolidation and Desalination of Ceramic Impregnated with Sodium Sulphate. MRS Proceedings, 267, 1071 doi:10.1557/PROC-267-1071.
- T. Y. Wu and D. L. Questad (1992). Mechanical Characterization and Constitutive Description of Thin-Film Polymeric Materials. MRS Proceedings, 264, 143 doi:10.1557/PROC-264-143.(Jr., n.d.)
- Jr., W. D. C. (n.d.). *Materials Science Engineering, An Introduction.* (third edit., p. 785).

DESMAR SOLAR

- Suzana P. Nunes, Ali Reza Behzad and Klaus-Viktor Peinemann (2013). Self-assembled block copolymer membranes: From basic research to large-scale manufacturing. Journal of Materials Research, 28, pp 2661-2665. doi:10.1557/jmr.2013.253.
- Paterakis, A. B. (1992). The Consolidation and Desalination of Ceramic Impregnated with Sodium Sulphate. *MRS Proceedings, 267.*
- Paul C. Painter, M. M. C. (1996). *Fundamentos de ciencia de polimeros: un texto introductorio* (p. 467). Cambridge University Press.
- Zalba, B. (2003). Review on thermal energy storage with phase change: materials, heat transfer analysis and applications. *Applied Thermal Engineering*, *23*(3), 251-283. doi:10.1016/S1359-4311(02)00192-8

DESMAR SOLAR

THANK YOU FOR YOUR ATTENTION

DESMAR SOLAR

The PSM eBook™

Bringing an Industrial Application to University Classrooms and Workplaces

AICHE New Orleans - 2014

The PSM eBook™

AICHE / CCPS New Orleans
31 March 2014

Process Safety Management (PSM) is a critical issue for both industry and academia. Delivering the required information and training in today's Internet oriented environment requires a creative approach to be efficient, effective and accepted by the computer oriented students, faculty and industry. ThePSMeBook™ delivers the user friendly learning tool that can be used as reference or refresher training in industry as well as a text book for engineering and technology students.

ThePSMeBook™ introduces Process Safety Management based on the US OSHA regulation, 29 CFR 1910.119. This is a very complex regulation. A very structured approach is required to link the subject areas into an understandable philosophy about how facilities should be operated safely. It should be noted that the regulation is performance oriented rather than prescriptive driven. Thus, the philosophy of why is of utmost importance. These practices extend across industry – refining, chemicals, exploration & production, nuclear, food industry as well as academic laboratories. While industrial manufacturing has a larger risk window, we have seen significant academic laboratory incidents in the recent past. Broad education in PSM practices not only makes graduates more attractive to industrial hiring companies, it also makes them safer in their laboratories and daily practices. ~~The~~ ThePSMeBook™ provides an efficient cost effective tool for PSM learning whether it is classroom, self-study, reference material or refresher training.

ThePSMeBook™ represents an innovation in learning delivery. It is not a simple PDF document with links in the table of contents to the chapters. This is an interactive tool that integrates voice over discussion and web-links to a wealth of Internet content. The material stays up to date as the internet updates information. The text of the eBook is also updated so that users can always have the most recent version with current internet content. Quizzes (essay and multiple-choice) and exams are built in and can be managed by the professor. He/she can receive student test results electronically after each quiz or exam, as well as how the students are progressing through their study of the material. The quizzes measure comprehension, and in-progress and outcome metrics can be collected at the professor's desktop.

These instructor-friendly features and metrics are key aspects of the Universal Technical Systems, Inc., or UTS Learning Center software. The UTS software can be installed on a university or corporate server, or on any Windows® personal computer or tablet. The business model for ThePSMeBook™ is a personal, university or corporate license, renewable annually. UTS is also the developer of the following technical software: TK Solver™, Roark's Formulas for Stress and Strain, Advanced Spring Design and Integrated Gear Software. All of these products are able to be integrated into ThePSMeBook™ platform when needed or required.

The PSM eBook™

Ronald L Cutshall Sr, PE, CEng   Deborah Grubbe, PE, CEng   Steven J Swanson, PhD, ChE
thePSMebook@comcast.net

The PSM eBook™

A significant positive feature of ThePSMeBook™ is that it delivers learning in a method that students and young professionals are very accustomed to and greatly prefer over paper text books that are far more expensive and yet unable to harness the power of the Web in real time.

PSM knowledge and skill is required both in industry as well as university laboratories. University laboratories have had several recent accidents that may have been prevented by a broader education in PSM for both students and faculty. Companies that are hiring far prefer and sometime require PSM training as a prerequisite for a job offer. PSM is a complex topic that requires detailed study to fully appreciate. For Chemical Engineers, it is a "form of art" that must be integrated into all aspects of their discipline. From the first twinkling in the eye of a research chemist to the ultimate design and startup of a new technology, the Chemical Engineer MUST integrate PSM into the total process. It is critical to embed this knowledge into the designers of our future manufacturing facilities. This knowledge is not generally taught specifically in industry. Graduates with a strong background in PSM are more valuable new employees and can get off to a quicker start. ThePSMeBook™ is a resource the student can use for years to come after graduation.

The base content for ThePSMeBook™ is a teaching slide pack used at Purdue University to teach a PSM course in Chemical Engineering for Juniors, Seniors and Graduate Students. The course was also opened to Mechanical Engineers. The materials contained in ThePSMeBook™ provide a basic requirement for innovators to meet before bringing a new invention to market. It gives the innovator the opportunity to develop greater credibility with industrial contacts. The materials provide a starting point for ALL instructors on integrating PSM into some aspect of their course materials. Feedback from industry clearly indicates that graduating undergraduates will greatly benefit from having this type of training on their transcript.

Practicing engineers have several ways they can utilize ThePSMeBook™. It is common for companies to have specialized training programs for new hires. The ThePSMeBook™ is a great resource for conducting such training. Additionally, refresher training is frequently required on critical topics. Given the included quizzes and tests, this tool could be used to meet this requirement. It can be integrated into existing testing systems. Experienced engineers occasionally need a PSM reference and ThePSMeBook™ is much more user friendly than trying to read the regulations.

The material for ThePSMeBook™ was developed for a 3 credit hour PSM Course for the Purdue School of Chemical Engineering. The class was taught twice as an elective before making it a required course for Chemical Engineers. The course is open to other engineering disciplines and some Mechanical Engineers have taken it. At the end of each class, the course is evaluated by the students. The class has been given exceptionally high marks by the students. 100% of the students stated that the PSM course was useful for their future. 100% of the students were better able to identify hazards both at home and at work. 100% of the students better understood PSM and how it is integrated into the work they would do in industry. Industrial contacts have given very positive feedback on the PSM knowledge of students who took the course.

The teaching materials for the course were used to create ThePSMeBook™. References, internet links, audio files, pictures and enhanced descriptions were added to the class materials.

The PSM eBook™

Additional chapters were added to broaden the scope of the eBook. Future editions will contain more international content. PSM elements from the 2007 text entitled "Risk Based Process Safety" will be added, as well as additional industrial examples, such as nuclear, LNG and Offshore Oil and Gas.

The Institute for Sustainability of AICHE is using ThePSMeBook™ for training to support its Sustainability Credential. A handful of universities and one large oil and gas producer are actively exploring, or have already installed ThePSMeBook™ software.

A live demonstration of ThePSMeBook™ via GoToMeeting is available upon request. A limited time usage can be arranged for those who would like to "test drive" the eBook. You can request more information by sending an e-mail to: thePSMeBook@comcast.net. Our web site is located at: http://www.uts.com/ThePSMeBook. A brief demo if the book is available here. We have a LinkedIn community (ThePSMeBook) and a twitter account, @ThePSMeBook.

Licenses are priced very reasonably and competitively. To order ThePSMeBook™ please visit this website: http://www.uts.com/OnLineCatalog.asp?ItemGroupID=2450 Because of the method of delivery, ThePSMeBook™ will be less expensive than printed texts. The book will be regularly updated. Please contact us with any questions at: thePSMeBook@comcast.net.

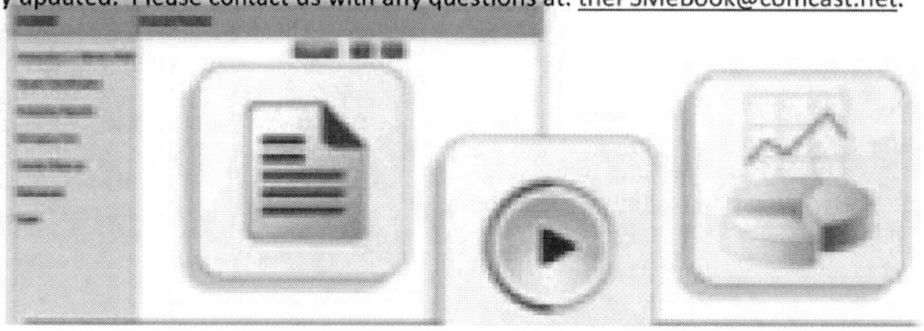

The PSM eBook™

Ronald L Cutshall Sr, PE, CEng Deborah Grubbe, PE, CEng Steven J Swanson, PhD, ChE

thePSMebook@comcast.net

347867 How to Start and Integrate into Industry after University

Abstract:

Transition from the well-known University environment into industry is not always easy and the expectations on both sides might be different, causing frustration. How can this be done more efficiently? What is an effective team composition, but most importantly what is a motivating environment for all employees, creating high performance teams. Consequently, what is effective leadership?

347838 Virtual Reality Technology for Knowledge Transfer in Process Facilities

Abstract:

Traditional chemical engineering has been undergoing several innovations related to market dynamics, especially in the oil and gas industry where the apparent sudden entry of natural gas is driving creative technology strategies to resolve resource, environment and safety topics. To summarize the problems, in addition to the growth in certain process engineering sectors there is a decrease in human resources as a result of the retiring work force and a dramatic increase in data as a by-product of complexity and innovation. As a result, one of the areas that has seen a lot of interest recently is interactive visualization of process facilities. Major companies are looking at this technology to facilitate digital retention of retiring knowledge as well as enable new engineers. Universities have also started to look at these technologies to train students on complex engineering topics.

This papers overviews the different interactive visualization technologies and their drivers. How are technologies such as virtual reality and augmented reality used to improve process facilities engineering? The paper also reviews when and how interactive visualization tools add value in the asset lifecycle. Finally this paper will present the maturity of these technologies in the industry and provide forward looking ideas of what is to come.

349166 Glycerol as a Raw Material for Lactic Acid Production for Industrial Process Course

Abstract:

Colombia is a palm oil producer, mainly for biodiesel by transesterification process, and the Glycerol is a sub product. Several Research Centers have been investigating in the possible uses for this raw material. The present Job is associated with the application of the contents of several tools used in industrial engineering curricula for the feasibility study for lactic acid production from glycerol. The application of the different principles studied in industrial Engineering Courses and the uses of computational tools, data bases and free software are shown in this proyect.

The Industrial Engineering program at UNIMINUTO is relatively new program in Colombia. The curricula have courses like Chemistry, Technical Drawing, Industrial process, thermodynamic, sustainable development and Economy, among others. In the present work, the feasibility of lactic acid production, different topics of these all materials were employed, and are described here. An idea was proposed, what can we do with the glycerol, from biodiesel production? And how obtain develop and prosperity from this resource? Is that possible? To answer these questions, a literature research was carried on. Several alternatives were found. In this part the chemistry was the support. After that, important economics concepts were used for make a decision of which products could be obtain and ultimately the concepts of industrial process and sustainable development were employed to obtain the final results.

The main idea is to shown how with this bases and these resources, a new process for the Acid Lactic using glycerol can be generated. With this can be seen how integrating engineering tools is possible answer critical questions about the education for new engineering generationSeveral software have been used in this project VISIO, AUTOCAD, EPIWEB.

The present research is classified as young researcher group in the Industrial Engineering Department in the UNIMINUTO.

Integrating K-12 Outreach & Service Learning with a Freshman Chemical Engineering Course

Bill B. Elmore, Ph.D., P.E.
Hunter Henry Chair & Interim Director

Swalm School of Chemical Engineering
Mississippi State University

Overview

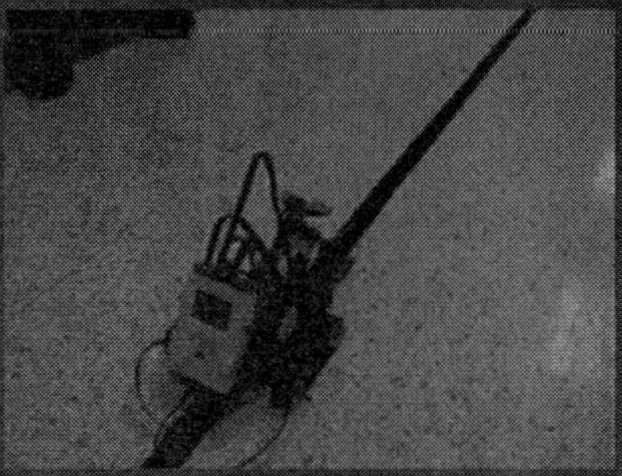

- Outcomes & Objectives
- Freshman Course Structure
- Benefits of using LEGOs
- Outreach development
- Strengths and weaknesses
- Future Goals

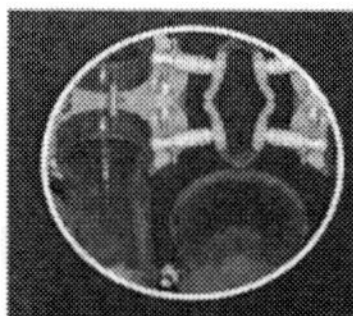

Student Outcomes

CHE 2990

By the conclusion of Intro to ChE, you should have...

- *understanding chem. engr problem solving (f, h, i, j, k)**
- *engr design, teamwork, study habits (i, k)*
- *sense of the chemical industry (f, h)*
- *understanding of the role of safety in the chemical industry (f, h, j)*
- *professional and ethical issues that arise in engineering practice (f, h, i)*
- *knowledge of our curriculum structure and career paths (i, k)*

*(letters in parentheses correlate to the ABET Accreditation criteria—Student Outcomes):

CHE 2213

By the conclusion of ChE Analysis, you should be able to...

- *Analyze & solve fundamental chemical engineering problems(a, b, c, e, k)*
- *Use Excel tools for making design and performance improvements. (a, b, c, g, I, k)*
- *Use Engineering Design Cycle (b, c, d, e, f, g, h, i, j, k)*
- *Work with team members (c, d, e, f, g, k)*
- *Develop a functional process from concept to improved design (a, b, c, d, e, f, g, k)*
- *Explain to someone in your family (or a non-engineer) what chemical engineering is about—giving some very practical examples (g h i, j)*

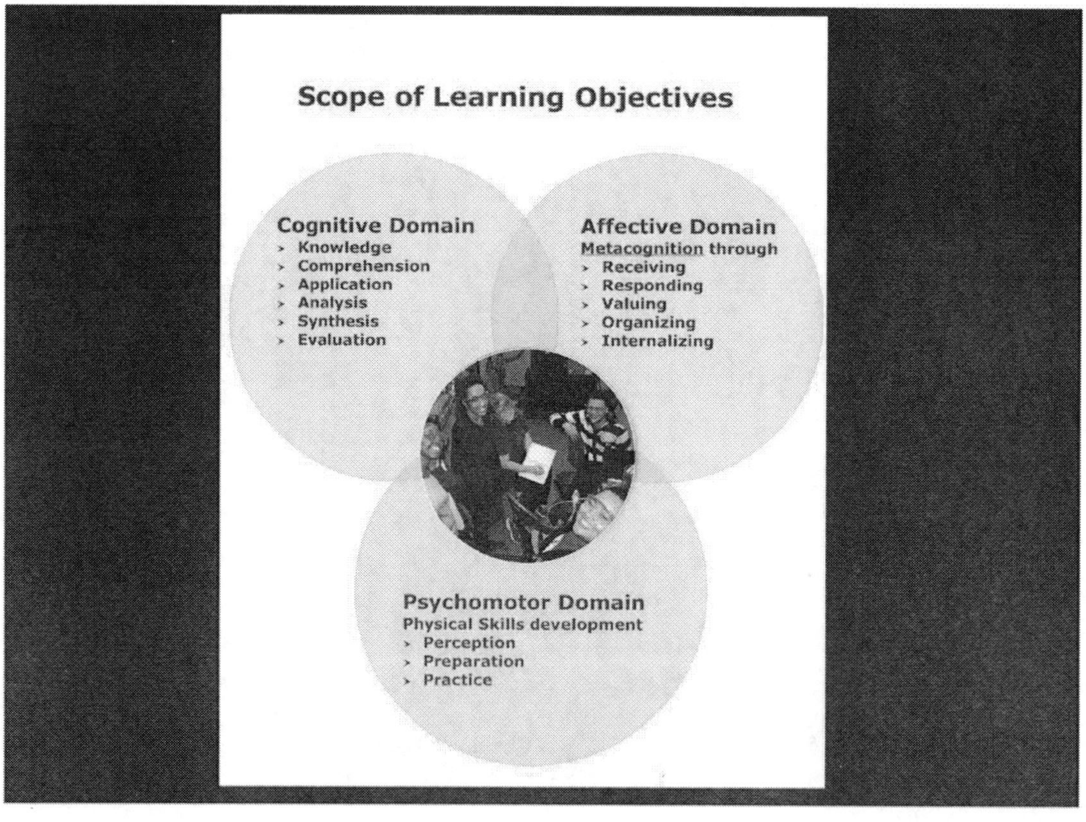

Team Challenges
Communication
• Teamwork
• Engineering logbooks
• Written & Oral reports
• Poster symposium
General Engineering & ChE -specific Topics
• Units/Measurements
• Material Balances
• Fluids/Heat Transfer
• Data collection & analysis
• Systems control
Engineering Analysis & Design
• Problem definition
• Develop prototype from promising design ideas
• Test, evaluate, improve
• Process Design & Economics
• Communicate "optimum"
Scope of Team Challenges

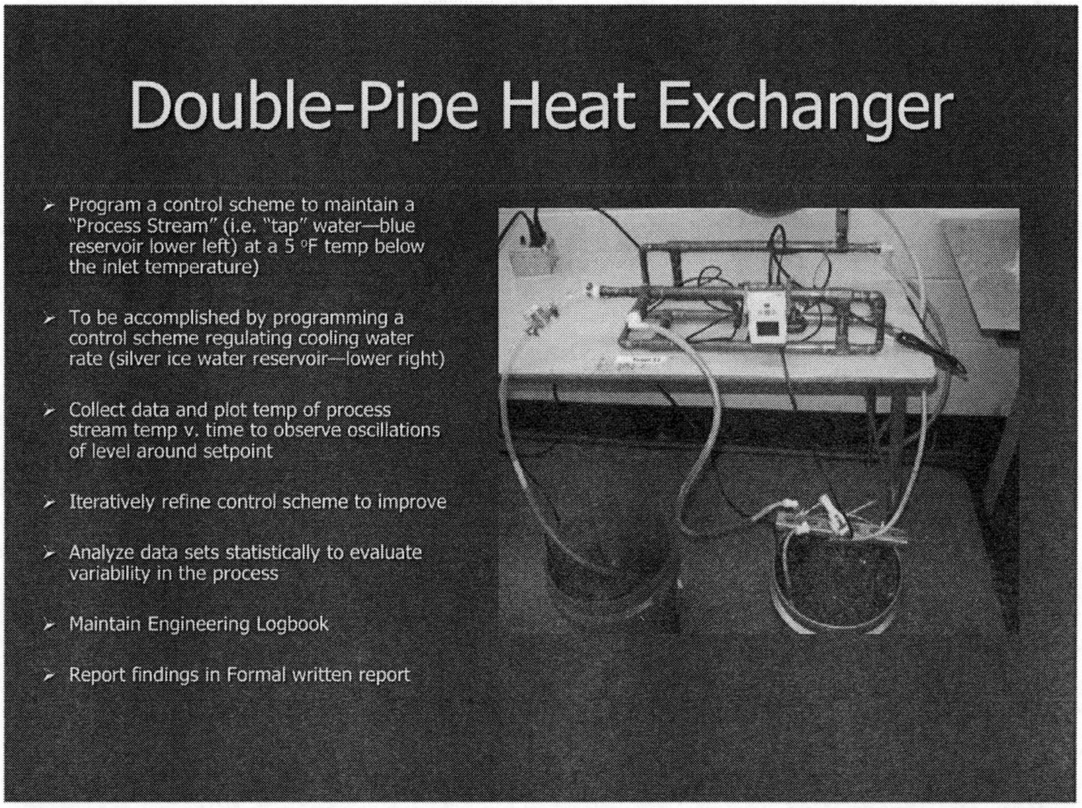
Double-Pipe Heat Exchanger
➢ Program a control scheme to maintain a "Process Stream" (i.e. "tap" water—blue reservoir lower left) at a 5 °F temp below the inlet temperature)
➢ To be accomplished by programming a control scheme regulating cooling water rate (silver ice water reservoir—lower right)
➢ Collect data and plot temp of process stream temp v. time to observe oscillations of level around setpoint
➢ Iteratively refine control scheme to improve
➢ Analyze data sets statistically to evaluate variability in the process
➢ Maintain Engineering Logbook
➢ Report findings in Formal written report

Tank Level Control

- Program a control scheme to maintain tank level at chosen setpoint
- Collect data for multiple iterations and plot level versus time to observe oscillations of level around setpoint
- Iteratively refine control scheme to improve
- Analyze data sets statistically to evaluate variability in the process
- Maintain records in Engineering Logbook
- Report finds in formal written report

Benefits of using LEGOs™

- Benefits
  - Familiar platform (LEGO bricks)
  - "Seamless" integration of hardware and software
  - Minimal startup delays
  - Relatively low cost per unit

Development of Service Learning

- "Day visits" to schools throughout Mississippi and at MSU
- Pre-assembled, pre-programmed robots to maximize use of class time
- Activities gauged to fit grade level (i.e. middle school v. high school)
- Students focus on design elements and process steps

Program Activities

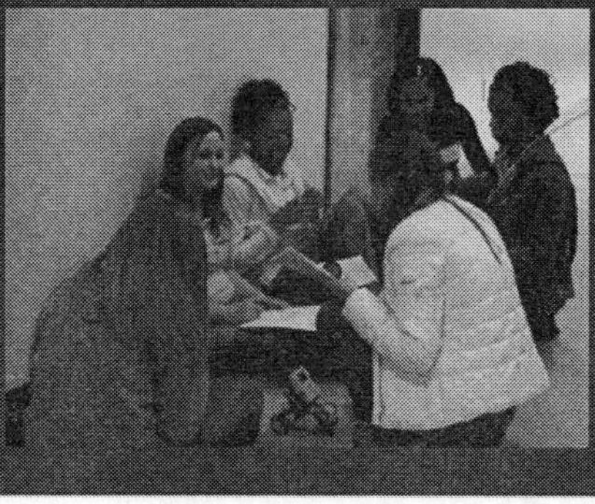

- Elementary
  - Discussion of how robots perform important tasks
  - Demonstrate simple automated tasks
  - Show integration with ChE concepts
- Middle School
  - Basic Engineering Concepts illustrated by pre-programmed tasks
  - Experimentation/data collection and analysis
- High School
  - Same as middle school with...
    - Robot assembly & programming

Recent Progress

- Outreach with Scouting (just finished a Girl Scout Merit Badge day with ~100 Scouts, ~30 Parents/Leaders, and 30+ AIChE/SWE Students
- Summer camps (Upward Bound-STEM prep for underrepresented groups/Jigsaw—outreach with autistic students/Quest—rising high school seniors)
- "WISE" workshops—Women In Science and Engineering

Strengths and Weaknesses

- Day visits provide broad exposure but limited interaction
- Sustained contact with individual schools offers greater impact
- Testing needed to evaluate program efficacy within proposed theoretical framework
- Sustained exposure with various outreach activities creating a working relationship with different groups.

Future Goals

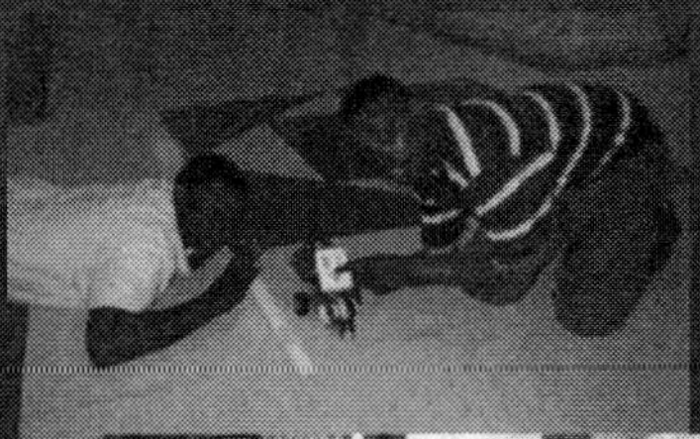

- Assessment of ongoing middle school program
- Increase interaction with pre-service teachers in College of Education
- Proposal refinement for teacher training & curriculum development
- Funding!

Questions?

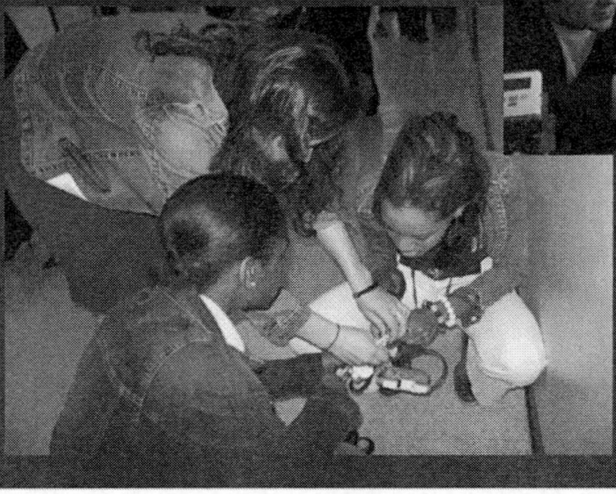

PROSYS™ Maximize Your Plant.

Operator Errors and What Can be Done to Minimize

Global Congress on Process Safety

March 31, 2014

10TH GLOBAL CONGRESS ON PROCESS SAFETY

PROSYS™ Maximize Your Plant.™

10TH GLOBAL CONGRESS ON PROCESS SAFETY

Presenter

Tom Nolan

- **Graduated from Ohio University with B.S. in Chemical Engineering**
- **24 years experience in chemical process industry in a variety of roles**
- **MAIC and DFSS Six Sigma Black Belt**

PROSYS Maximize Your Plant.™

10th GLOBAL CONGRESS ON PROCESS SAFETY

The Cost of Errors

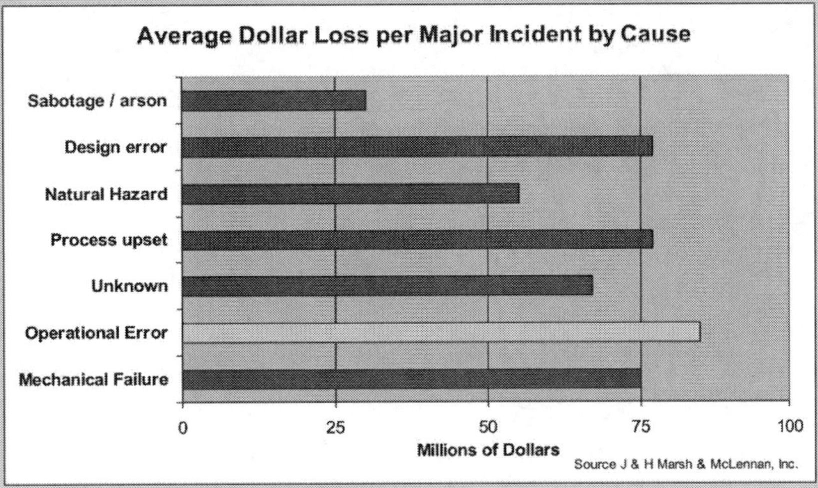

PROSYS Maximize Your Plant.™

10th GLOBAL CONGRESS ON PROCESS SAFETY

The Cost of Operator Errors

- **ASM estimates total loss due to operator error is $8B per year**
- **Chemical Safety Topical Committee - average of one chemical incident per day - cost of over $2 million per incident to comply with requirements**
- **Errors cause 42% of unscheduled shutdowns**
- **70% of process incidents occur during start-up or shutdown**

PROSYS Maximize Your Plant.™

10th GLOBAL CONGRESS ON PROCESS SAFETY

Reasons for Errors

- Lack of Skill
- Lack of Knowledge
- Carelessness
- System Design
  - Operator set up to make errors by inappropriate design – built in errors
  - Errors that are predictable are preventable by better design

PROSYS Maximize Your Plant.™

10th GLOBAL CONGRESS ON PROCESS SAFETY

Focus of this Discussion

- System Design
  - Alarm Management
  - Operator Graphics
  - Operator set up to make errors by inappropriate design – built in errors
  - Errors that are predictable are preventable by better design

PROSYS Maximize Your Plant™

Alarm Floods

In a number of industrial incidents, alarm floods were identified as a significant contributing cause to the incident…

As found by EEMUA in 1999 and CSB

Alarm Flood defined by ISA 18.2 as -
"10 or more annunciated alarms in any 10 minute period per operator"

PROSYS Maximize Your Plant™

Why do Alarm Floods Occur?

One reason is not providing dynamic alarm management

- **Alarms need to indicate abnormal situations that require operator action**
- **Processes do not operate in one state**
- **What is normal vs abnormal changes with state**
- **Alarms are typically configured for run therefore many alarms are triggered upon a Change of State: Run to Shutdown**
- **Many of the alarms are not applicable or actionable for the new state impleading the operators ability to act quickly on what is important**

PROSYS Maximize Your Plant.™

What is Impacted by Alarm Floods?

- **Product quality**
- **Operability or profitability of the process**
- **Loss of equipment**
- **Loss of containment – environmental releases**
- **Injury and loss of life in plant or community**

PROSYS Maximize Your Plant.™

What Makes Alarm Floods so Dangerous?

Can be a problem for three reasons:

- **A deluge of alarms can cause critical alarms to be missed**
- **Floods can be a significant distraction when dealing with process upsets**
- **Can be an indicator of larger systemic safety issues**

PROSYS Maximize Your Plant.™

10th GLOBAL CONGRESS ON PROCESS SAFETY

Impacting Alarm Management Design

- **Alarm rationalization is not a process to eliminate alarms - its about quality**
- **Good rationalization will add alarms when appropriate**
- **Typical (Static) rationalization is only for run mode**
- **Dynamic rationalization considers all plant modes**

PROSYS Maximize Your Plant.™

10th GLOBAL CONGRESS ON PROCESS SAFETY

Dynamic Alarm Management

- **Dynamic rationalization does everything a static rationalization does plus asks "when" for each alarm**
- **Requires dynamic software to make changes based upon operating mode of the plant**
- **Eliminates redundant alarms and lowers operator loading during transitions**
- **Only alarms what is abnormal and actionable for the given state**

PROSYS Maximize Your Plant.™

10th GLOBAL CONGRESS ON PROCESS SAFETY

Actual Performance Metrics vs ISA 18.2

Type	Avg Alarm Rate/hr	Low/Hi of Avg Rate/hr	Peak Alarm Rate/hr	Low/Hi of Peak Rate/hr	% Time in Flood	Standing Alarms >24 hrs
Before Rationalization	30	Low=6.2 Hi=61	638	Low=152 Hi=2402	17.2%	9
ISA 18.2 Target Metrics	6	---	<60	---	<1%	<5
After Dynamic Alarm Management	2	Low=0.09 Hi=2.5	25	Low=13 Hi=42	0.25%	4

PROSYS Maximize Your Plant.™

10th GLOBAL CONGRESS ON PROCESS SAFETY

Results After Dynamic Alarm Management

Annunciated Alarms Per Hour (Monthly) **Area: G1-G5-G6** **June 2013**

Average Counts | Peak Counts | Total Counts

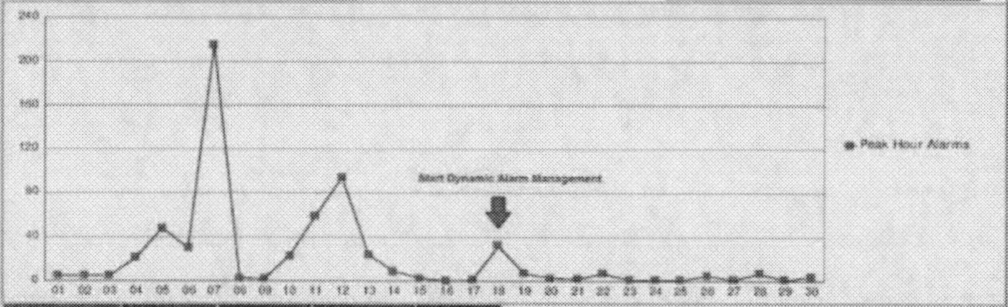

Day	Min Hour Alarms	Avg Hour Alarms	Peak Hour Alarms	Total Alarms
01-Jun-2013	0	0.3	5	6
02-Jun-2013	0	0.4	5	9
03-Jun-2013	0	0.5	5	13
04-Jun-2013	0	1.1	21	27
05-Jun-2013	0	13.3	48	319
06-Jun-2013	0	3.6	30	86
07-Jun-2013	0	11.9	215	285
08-Jun-2013	0	0.4	3	10
09-Jun-2013	0	0.3	2	7
10-Jun-2013	0	1.5	23	35
11-Jun-2013	0	4.7	59	112
12-Jun-2013	0	11.9	94	285
13-Jun-2013	0	3.8	24	90
14-Jun-2013	0	1.3	9	32
15-Jun-2013	0	0.3	3	8

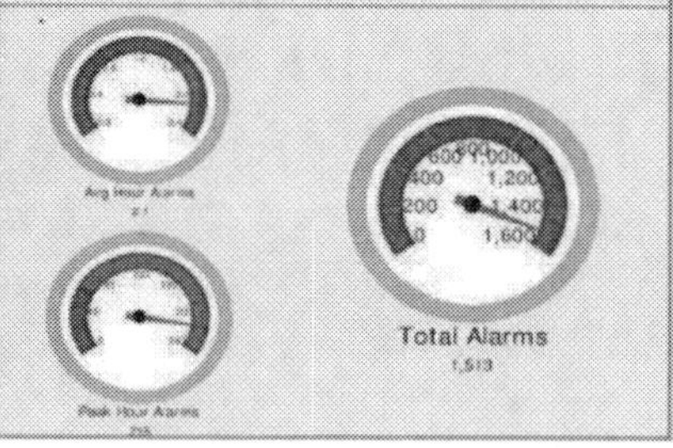

PROSYS™ Maximize Your Plant™

Operator Graphics – The Risk of Changing Focus

- **Distraction of changing focus from process graphic to faceplate window**
- **Added workload from managing multiple open windows**
- **Increased probability of errors when changing values for un-intended tag with multiple faceplates open**
- **All of above become multiplied and more complex when process is transitioning from one state to another**

PROSYS™ Maximize Your Plant™

Faceplate Operation

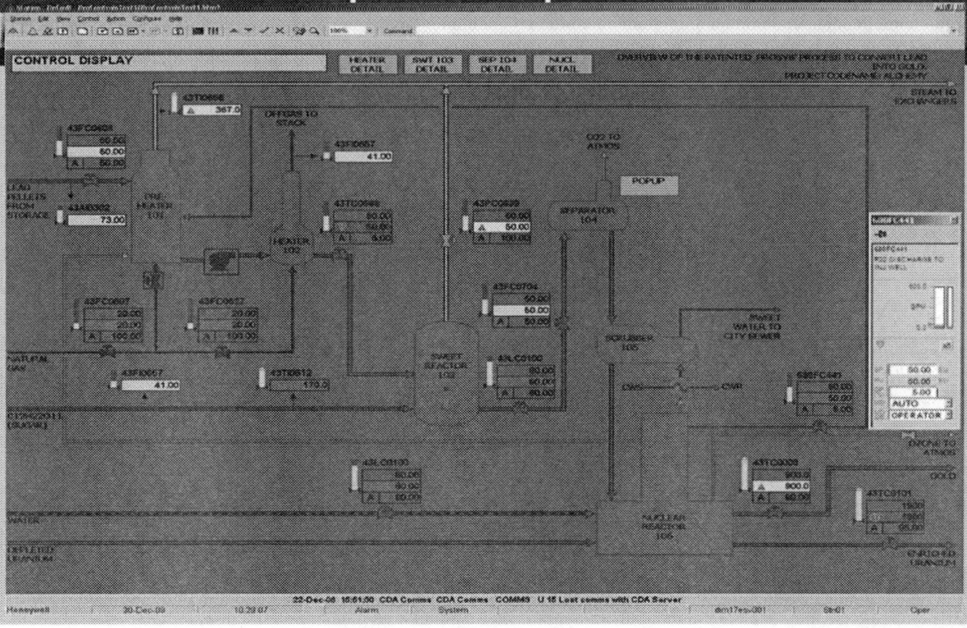

PROSYS Maximize Your Plant.™

10th GLOBAL CONGRESS ON PROCESS SAFETY

Potential for Errors

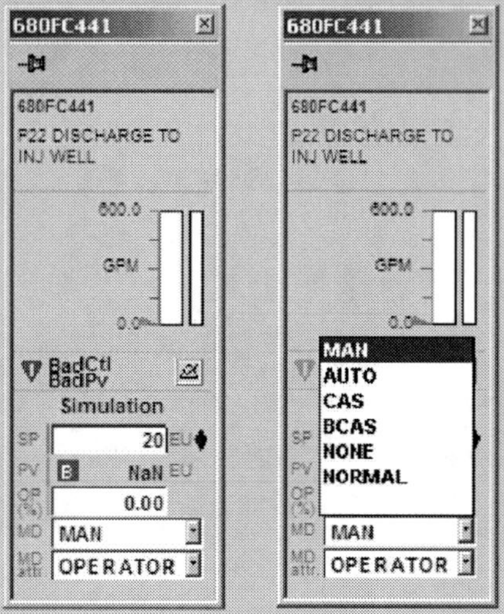

Faceplate Design -can introduce additional errors

- Mode drop down list covers SP, PV and OP values
- Operator may select wrong Mode if SP, PV and OP values are hidden from view
- Mode list offers more options than necessary

PROSYS Maximize Your Plant.™

10th GLOBAL CONGRESS ON PROCESS SAFETY

Direct Entry Operation

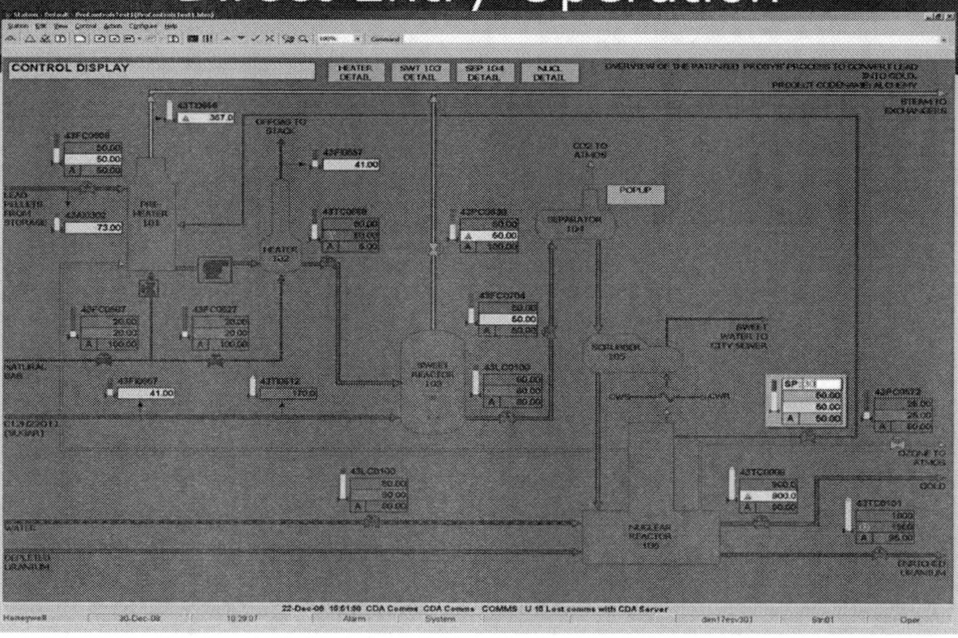

PROSYS Maximize Your Plant.™

Direct Entry Fields

Values can be entered by:

- Select point
- Typing value via keypad
- Pressing enter

Key Factors:

- By maintaining focus of points after value change, it improves safety for input errors, multiple clicking functions and is more time efficient
- Easily recognizable operator enterable fields

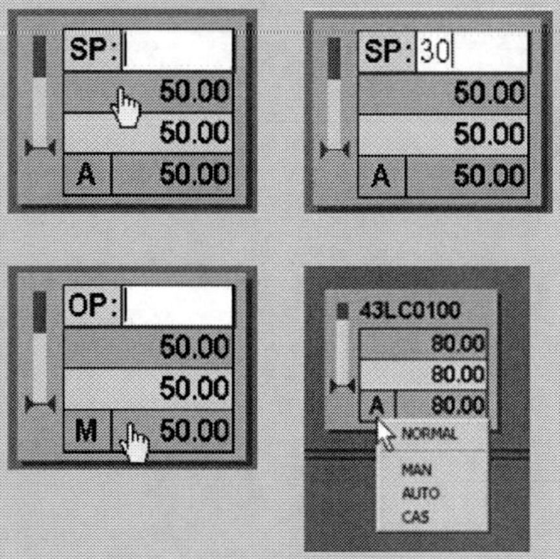

PROSYS Maximize Your Plant.™

Evaluation of Methods

Comparison of Faceplate to Direct Entry

- **Keystroke Level Analysis**
- **Operator Loading Analysis**
- **Risk Analysis**

PROSYS Maximize Your Plant.™

Keystroke Level Model

Symbol	Time (s)	Description
K	0.28	Keystroke
P	1.1	Point to object
BB	0.2	Click on object
H	0.4	Home hands on keyboard or mouse
M	1.2+	Mental act or routine thinking

PROSYS Maximize Your Plant.™

Estimate Execution Time Analysis

- **Evaluates execution time by an operator comparing faceplates to direct entry for SP/OP and Mode Changes**

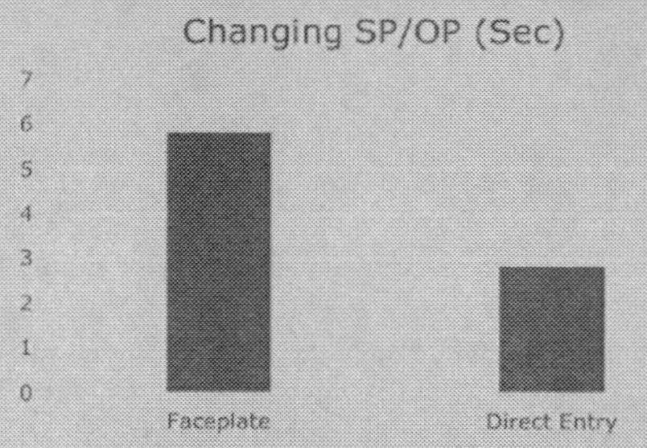

- **50 % Reduction**

PROSYS™ Maximize Your Plant™

Estimated Operator Loading Analysis

- **Evaluates Mental and Physical Load on Operator**
  - Tasks that require very little thought are removed – such as (BB), (H)

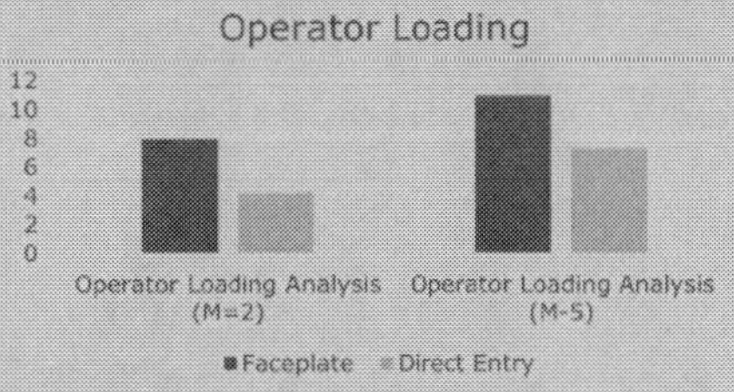

  - Mental and Physical tasks are weighted based on the amount of fatigue they cause – assigned a weight of M=2.0 to 5.0

PROSYS™ Maximize Your Plant™

Risk Analysis

- **Identifies actions where operator entry errors can occur**

PROSYS Maximize Your Plant.™

10th GLOBAL CONGRESS ON PROCESS SAFETY

Risk Analysis

Faceplate operation KLM for changing SP/OP	Potential Error
Move to shape (P)	Select wrong point
Click on shape (BB)	
Move to faceplate (P)	Point in faceplate is previous point
Click on SP/OP field (BB)	Change wrong parameter
Move hands to keyboard (H)	
Type in value and press enter (4K)	Mistype value and press enter
Move hands to mouse (H)	
Move to close faceplate (P)	
Click to close faceplate (BB)	
Potential error count	**4**

PROSYS Maximize Your Plant.™

10th GLOBAL CONGRESS ON PROCESS SAFETY

Risk Analysis

Direct Entry operation KLM for changing SP/OP	Potential Error
Move to SP/OP (P)	Select wrong parameter or point
Click on SP/OP field (BB)	
Move hands to keyboard (H)	
Type in value and press enter (4K)	Mistype value and press enter
Potential error count	**2**

PROSYS Maximize Your Plant.™

10th GLOBAL CONGRESS ON PROCESS SAFETY

Direct Entry Vs Faceplate

Summary of Benefits

- **Execution Time SP/OP Changes - 51.5% less**
- **Execution Time for Mode Changes – 50% less**
- **Operator Loading – 33 % to 47% less**
- **Risk of Errors – 50% less**

PROSYS Maximize Your Plant.™

10th GLOBAL CONGRESS ON PROCESS SAFETY

Conclusion

- **Many industrial errors are a result of operators using control systems with flawed designs**
  - Poor design and performance by alarm management distracts operators and/or occludes critical alarms
  - Operator graphics with multiple open faceplates can cause a change intended for one controller to be entered into the faceplate of another
- **The two mechanisms listed above are predictable and therefore preventable through better design available today**

PROSYS™ Maximize Your Plant.

visium KMS
Process Safety Management Solution
When Sharing is not Enough… HUAA
10th Global Congress on Process Safety – New Orleans, LA
Expert Insight Delivered
© 2014 Optimized Systems and Solutions Inc. All rights reserved.
Osys
Part of the Rolls-Royce Group

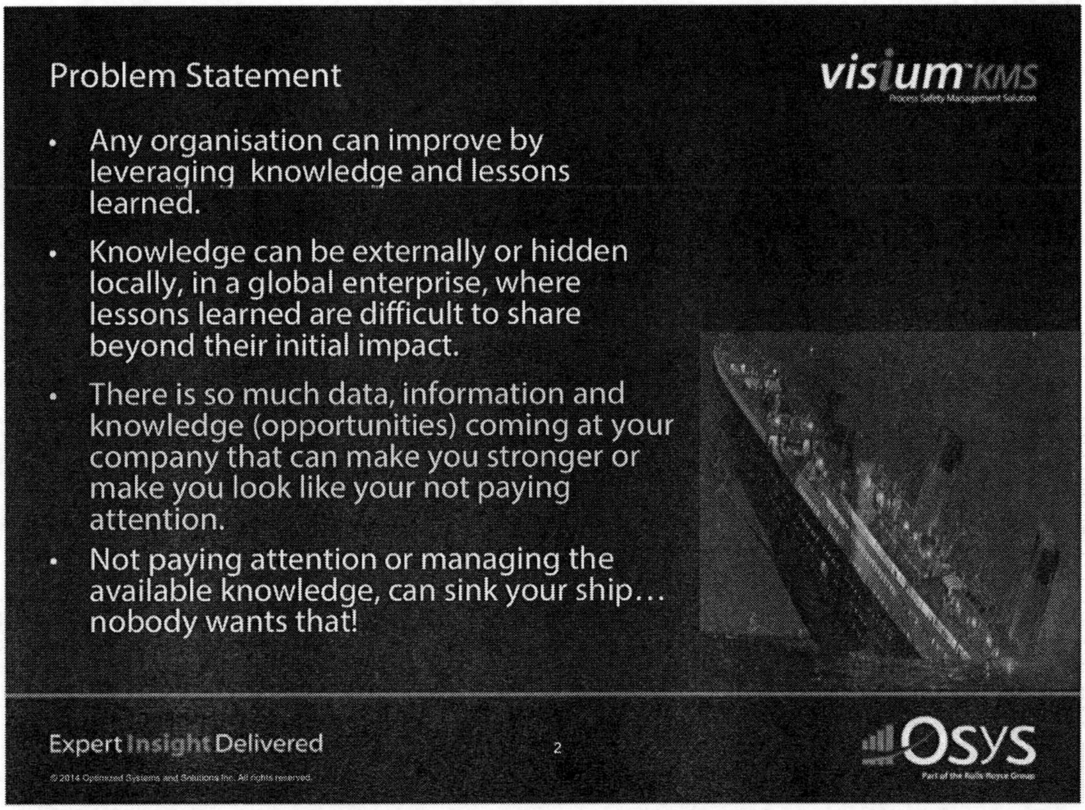
Problem Statement
visium KMS
Process Safety Management Solution
• Any organisation can improve by leveraging knowledge and lessons learned.
• Knowledge can be externally or hidden locally, in a global enterprise, where lessons learned are difficult to share beyond their initial impact.
• There is so much data, information and knowledge (opportunities) coming at your company that can make you stronger or make you look like your not paying attention.
• Not paying attention or managing the available knowledge, can sink your ship… nobody wants that!
Expert Insight Delivered
2
© 2014 Optimized Systems and Solutions Inc. All rights reserved.
Osys
Part of the Rolls-Royce Group

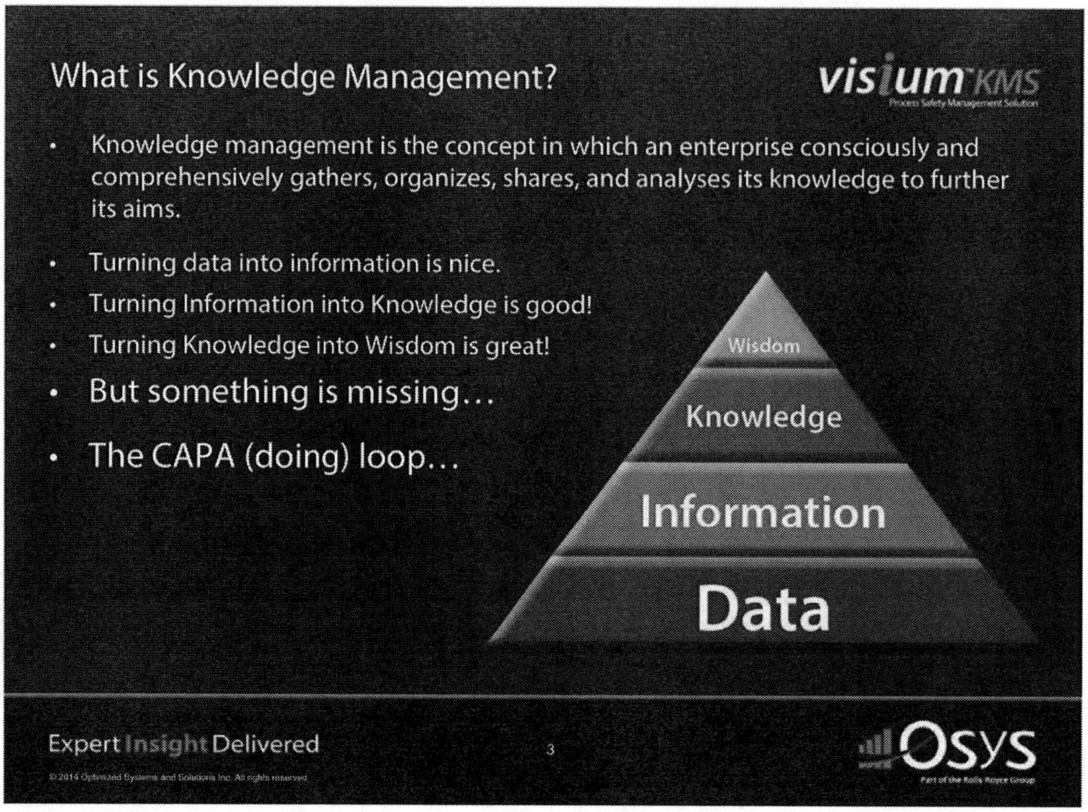
What is Knowledge Management?
visium KMS
Process Safety Management Solution
• Knowledge management is the concept in which an enterprise consciously and comprehensively gathers, organizes, shares, and analyses its knowledge to further its aims.
• Turning data into information is nice.
• Turning Information into Knowledge is good!
• Turning Knowledge into Wisdom is great!
• But something is missing…
• The CAPA (doing) loop…
Wisdom
Knowledge
Information
Data
Expert Insight Delivered
3
Osys
Part of the Rolls Royce Group

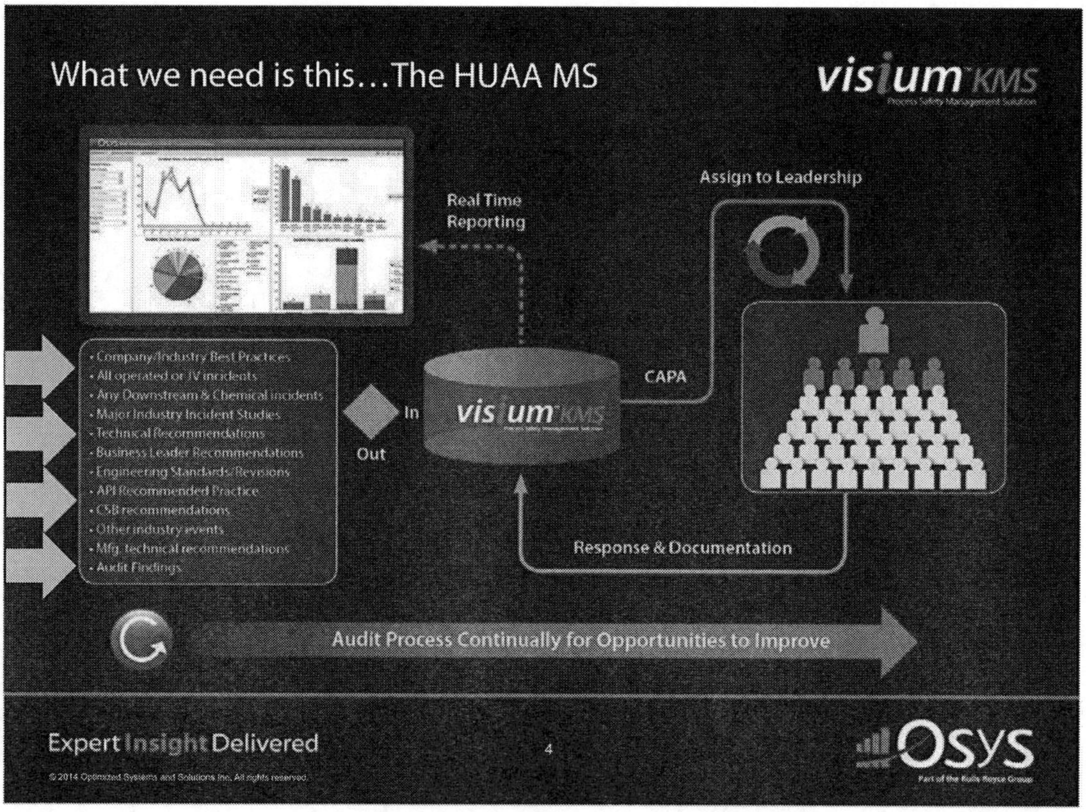
What we need is this…The HUAA MS
visium KMS
Process Safety Management Solution
Real Time Reporting
Assign to Leadership
• Company/Industry Best Practices
• All operated or JV incidents
• Any Downstream & Chemical incidents
• Major Industry Incident Studies
• Technical Recommendations
• Business Leader Recommendations
• Engineering Standards/Revisions
• API Recommended Practice
• CSB recommendations
• Other industry events
• Mfg. technical recommendations
• Audit Findings
In
Out
visium KMS
CAPA
Response & Documentation
Audit Process Continually for Opportunities to Improve
Expert Insight Delivered
4
Osys
Part of the Rolls Royce Group

What is the HUAA Management System…

visium KMS
Process Safety Management Solution

HUAA – Heard, Understood, Acknowledged, Actioned.

1. Opportunities identified (heard) by your experts,
2. Captured by an electronic system,
3. Expertly Reviewed (understood),
4. Accepted or rejected (acknowledged) and
5. Assigned (actioned) to Sr. Managers for resolution and tracked to closure.

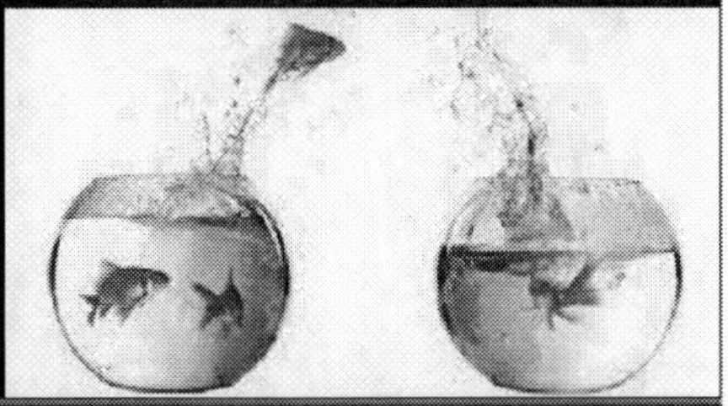

Expert Insight Delivered 5

Osys
Part of the Rolls-Royce Group

HUAA – Management Systems - Detail

visium KMS
Process Safety Management Solution

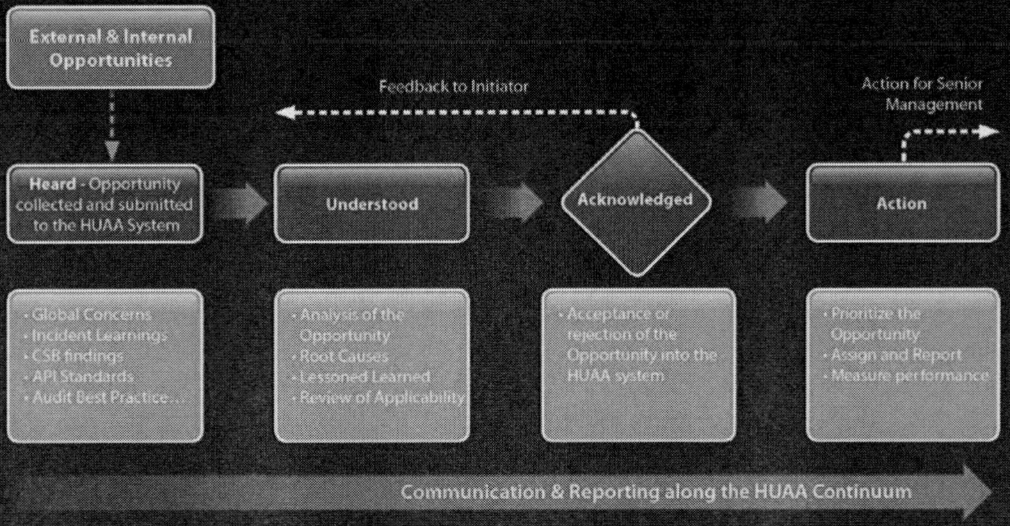

Expert Insight Delivered 6

Osys
Part of the Rolls-Royce Group

HUAA Results

visium KMS
Process Safety Management Solution

- Efficient Collection - Important Opportunities are identified and collected from many sources (internal and external), and entered into the HUAA management system.
- Expert Vetting - These opportunities are vetted, parsed and resolved for inclusion or exclusion; documenting & communicating both.
- Accountable Assignment - Included opportunities are systematically assigned to Sr. Leaders and managed across the enterprise at SVP and C levels.
- Action - Improvement will be assured through the accountable actions of these leaders.
- Results - The organization will gain competitive advantage by active listening, learning from their experience, as well as the experience of others, and by taking ACTION!

Expert Insight Delivered 7 Osys
Part of the Rolls-Royce Group

Sharing is not enough to keep your ship afloat!

visium KMS
Process Safety Management Solution

Expert Insight Delivered 8 Osys
Part of the Rolls-Royce Group

© 2014 Optimized Systems and Solutions Inc. All rights reserved.

352631 Autocad - Smart Grid – Agriculture - Architecture and the United States Government

Abstract:

The computational methods, process control, process design, process safety, public policy of the smart chip for agriculture industry, and architecture industry in the United States utilizing the computational methods of AutoCAD in the structural development stages of the smart grid for device/component/equipment and geographical development and public policy for grant funding.

352747 Production, Characterization and Catalytic Studies of Biobased Carbon Materials

Abstract:

Carbon-based nano materials were produced from wood chips, lignin and bio-char. The influence of reaction conditions on the production of nano-carbons was investigated. Carbon-based nanomaterials were characterized using scanning electron microscopy (SEM), transmission electron microscopy (TEM), high resolution transmission electron microscopy (HRTEM), and X-ray diffraction (XRD). The catalytic performance of these nanomaterials as syngas cleaning adsorbents and chemical conversion catalysts for bio-fuel processes has been evaluated.

347835 Uncertainty in Sour Gas Viscosities Estimation, What Is the Effect on Your Reservoir Inflow and Tubing Performance

Abstract:

In light of the need to find an applicable, consistent and promising correlation to accurately calculating sour gases viscosity under several reservoir conditions, this study aims to compare the existing correlations and other industry standard methods, check their accuracy and range of applicability for sour/acidic gases. Moreover, investigating the impact of uncertainties in estimating sour gas viscosities on reservoir flow performance relationship (IPR), reserves and resources estimations, tubing performance, pipeline and facilities design, the design of compressors and more importantly the need to find such correlation to be used in further advancements of reservoir simulation.

354052 Chromonic Nanocarriers for Chemotherapeutics: Size Distribution and Control Release Studies

Abstract:

Conventional chemotherapy fails to release the drug in the controlled manner; rather they release the drug instantly due to which re-administration is required as therapeutic index falls. Moreover, they target both the cancer cells and non-cancer cells which make the treatment of the cancer cells highly ineffective due to excessive toxicities. Recently, the nanoparticle mediated drug delivery for chemotherapeutics has gained much attention. The present study reports the use of a novel class of materials – *Chromonics* in cancer drug delivery. Folic acid as a chromonic material has been explored to develop nanocarrier for sustained release of chemotherapeutic drugs *in-vitro*. With the help of a biocompatible polymer as a tool, folate nano-domains were formed from self-assembled liquid crystalline folate solutions. These nano-domains were crosslinked to form a stable nanoparticle system. The effect of different factors like crosslinking agent, relative polymer and folate concentration, on the size distribution was investigated in detail. Based on the observations, we have reported that not only the synthesis but also the size of the developed nanoparticles can be controlled by varying the folate: HPMC weight ratios. The present study then explored the release of drugs from these nanocarriers. We have investigated the release of folic acid as a model drug from these nanoparticles as well as release of methotrexate and cytarabine as an anticancer drug for the period of 60 days. It was observed that release of drug was nanoparticle's size-dependent, release medium dependent, pH of the release medium dependent, crosslinking agent and its concentration dependent. The significant part of the current study is the synthesis of chromonic nanoparticles, how the size can be controlled, how controlled release can be obtained as well as how the release profiles can be modified depending on the dosage forms. Therefore, we can say that folate nanoparticles can be a potent nanocarrier for the sustained release of cancer drugs.

Double Containment Piping Systems: A Fail-Safe, Leak-Free Solution

IPEX
Committed to Excellence

Darin Johnson, P.Eng.
Applications Engineer
IPEX USA LLC

AIChE – 2014 Spring Meeting
New Orleans, LA

Monday, March 31, 2014

Why Double Containment Piping Systems?

IPEX
Committed to Excellence

- December 1988 – law change within the ***Resource, Conversation, and Recovery Act*** in the United States regarding underground piping.
- U.S. Environmental Protection Agency (EPA) requirement: Code of Federal Regulations (CFR), Title 40, Part 280.
  - Petroleum Products
  - Hazardous Substances (over 3200, CFR Title 40 Part 302.4)
  - Hazardous Wastes

Why Double Containment Piping Systems?

- Buried tank - all buried piping connected to it must be double contained.
- Above ground tank – associated buried piping consists of 10% or more of the tank volume, it also must be double contained.

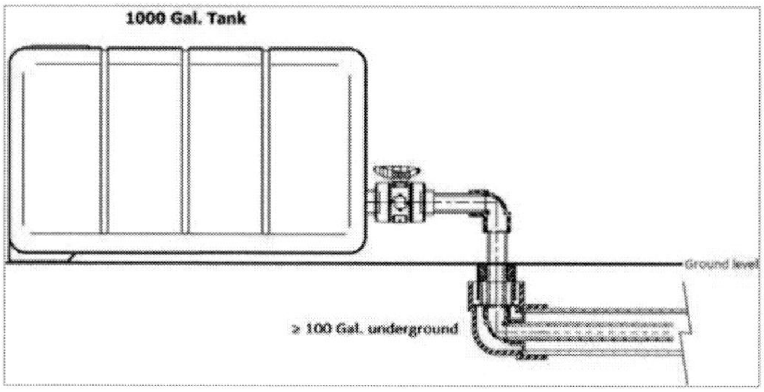

Why Double Containment Piping Systems?

- If the piping is less than 10% of the volume of the aboveground tank, and you choose not to double contain the buried piping, ***responsibility for any clean-up or remediation resulting from a release will still fall on the Client.***

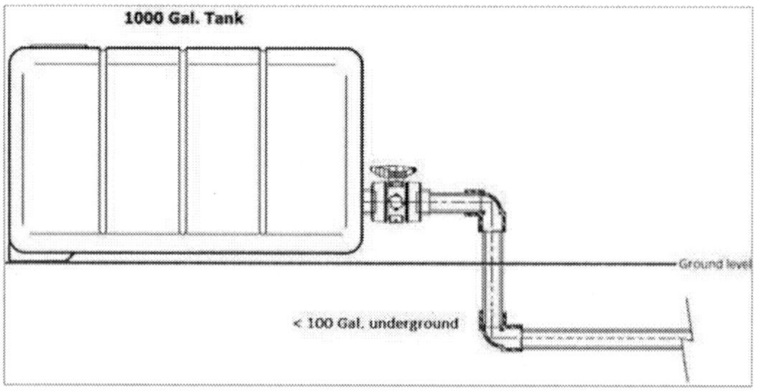

Why Double Containment Piping Systems?

- **Above Ground Piping Systems**
  - Different jurisdictions have to ability to require chemicals be double contained.

Nomenclature of Double Containment Piping

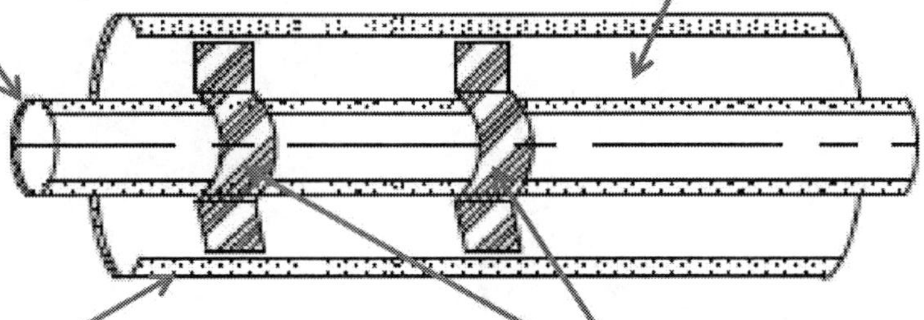

Double Containment Piping Systems: Material Combinations

SYSTEM Carrier	Containment	Carrier SIZE inches
PVC 80	PVC 80	1/2 - 12
PVC 80	PVC 40	1/2 - 12
PVC 40	PVC 40	1/2 - 12
PVC DWV	PVC DWV	1-1/2 - 12
PVC 80	CS SCH 10	1/2 - 12
CPVC 80	CPVC 80	1/2 - 8
CPVC 80	PVC 80	1/2 - 12
CPVC 80	PVC 40	1/2 - 12
CPVC 80	CS SCH 10	1/2 - 12
NAT PP 80	PVC 80	1/2 - 2
NAT PP 80	PVC 40	1/2 - 2
RED PVDF 80	CPVC 80	1/2 - 2
RED PVDF 80	PVC 80	1/2 - 2
NAT PVDF 80	PVC 80	1/2 - 2
ENFUSION PP	PVC DWV	1/2 - 6
S40 A53B CS BW	PVC 80	1/2 - 8
S40 A106B CS BW	PVC 80	1/2 - 8

SYSTEM Carrier	Containment	Carrier SIZE inches
S40 A53 BLK SW	FRP	1/2 - 2
S40 A53 BLK SW	PVC 80	1/2 - 2
S80 A53B CS SW	PVC 80	1/2 - 2
S10 T304L SS BW	PVC 80	1/2 - 8
S10 T304L SS BW	PVC 40	1/2 - 8
S10 T0304L SS BW	CPVC 80	1/2 - 8
S10 T304L SS BW	FRP	1/2 - 12
S10 T304L SS BW	S10 T304L	1/2 - 12
S40 T316L SS BW	PVC 80	1/2 - 8
S40 T316L SS BW	CPVC 80	1/2 - 8
S40 T316L SS BW	FRP	1/2 - 12
S40 T304L SS SW	PVC 80	1/2 - 2
S40 T304L SS SW	PVC 40	1/2 - 2
S40 T304L SS SW	CPVC 80	1/2 - 2
S40 T304L SS SW	FRP	1/2 - 2
S10 T316L SS SW	PVC 80	1/2 - 2
S10 T316L SS SW	CPVC 80	1/2 - 2

Double Containment Piping Systems: Design Considerations

- **Material Selection**
  - Primary Pipe
    - Chemical Compatibility
    - Pressure Capabilities
    - Thermal Properties
  - Secondary Pipe
    - Chemical Compatibility
    - Loading (Static & Live) for buried applications
    - Hanging requirements for above ground
    - Leak detection requirements
    - UV requirements

Double Containment Piping Systems: Design Considerations

- **Thermal Expansion & Contraction**
  - Two basic thermal relationships may be present:
    - When only one of the two systems experiences thermal expansion/contraction.
    - When both systems experience different magnitudes of thermal expansion/contraction.
  - The Floating System
  - The Restrained System

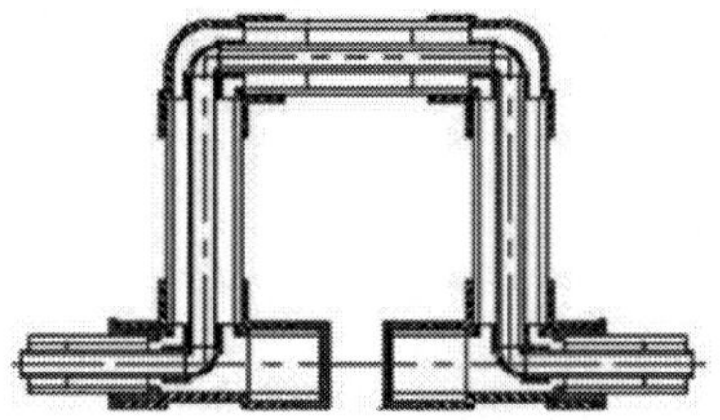

Double Containment Piping Systems: Leak Detection Requirements

- EPA requirement: (CFR), Title 40, Part 280
  - Pressurized systems require electronic leak detection that must be checked once per year.
  - Drainage systems (gravity flow) require visual leak detection with monthly manual inspection for a media leak.

IPEX
Committed to Excellence

Double Containment Piping Systems: Leak Detection Requirements

- The following forms of leak detection are acceptable as per regulations
  - Visual (drainage applications only)
  - Electronic low point probes
  - Continuous cable

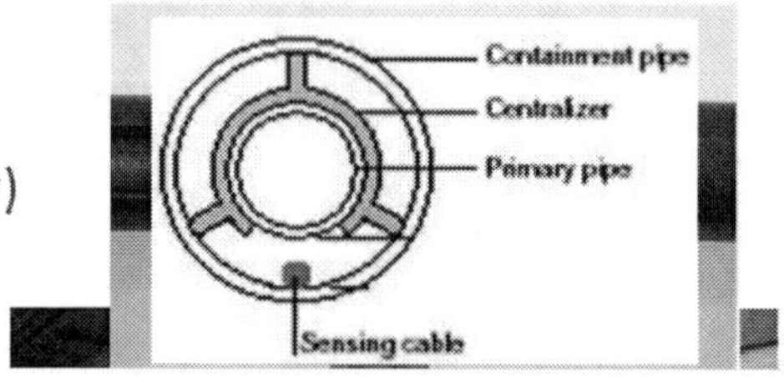

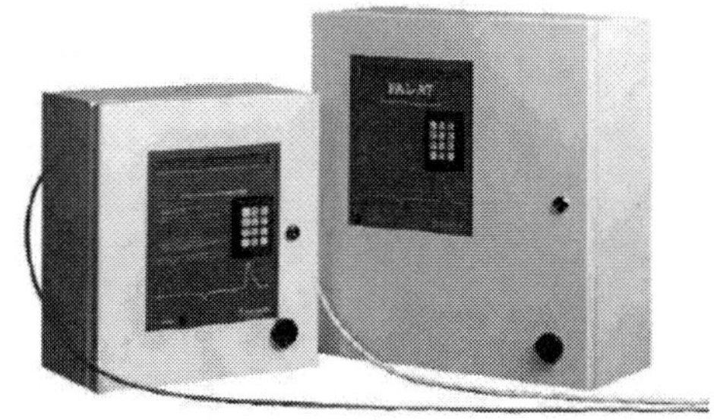

IPEX
Committed to Excellence

Double Containment Piping Systems: Applications

- Underground Applications
- WTP & WWTP
- Oil & Gas
- Hospitals
- Museums
- Food and Beverage
- Pharmaceuticals
- Laboratories
- Data Centers

Double Containment Piping Systems: A Fail-Safe, Leak-Free Solution

IPEX
Committed to Excellence

Questions?

Thank you!

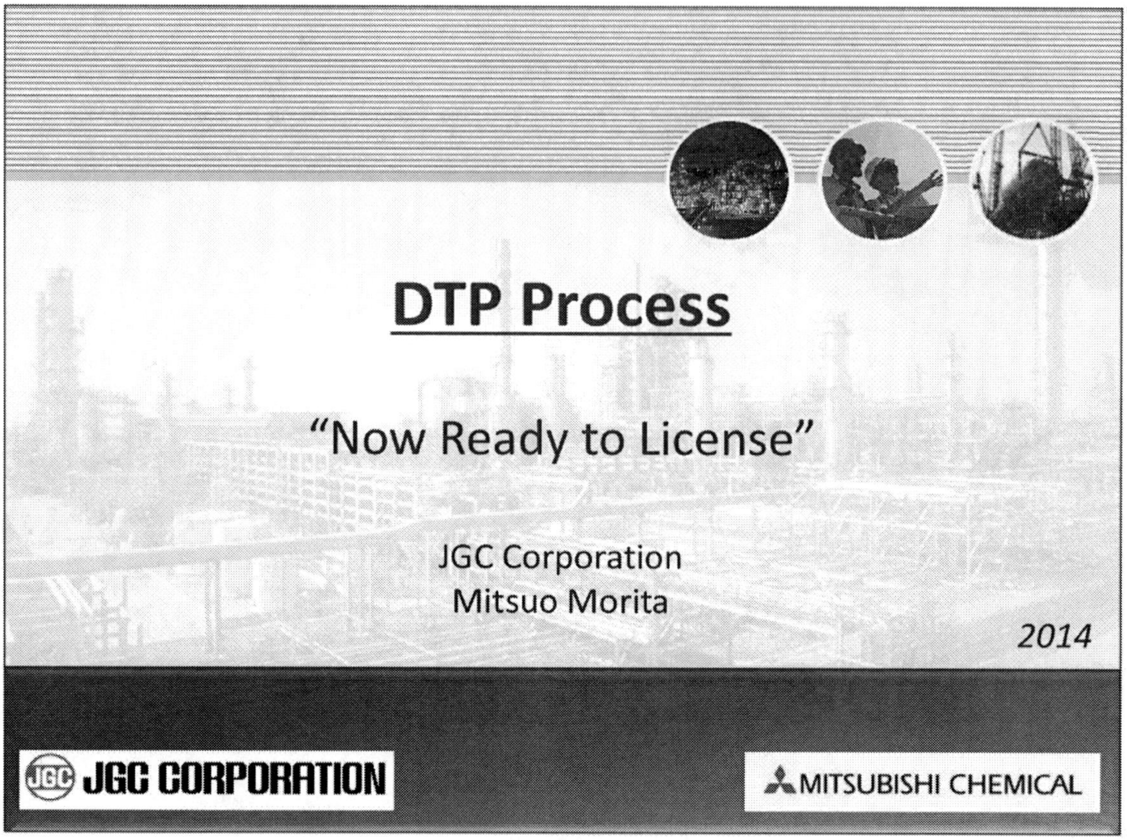
DTP Process
"Now Ready to License"
JGC Corporation
Mitsuo Morita
2014
JGC CORPORATION
MITSUBISHI CHEMICAL

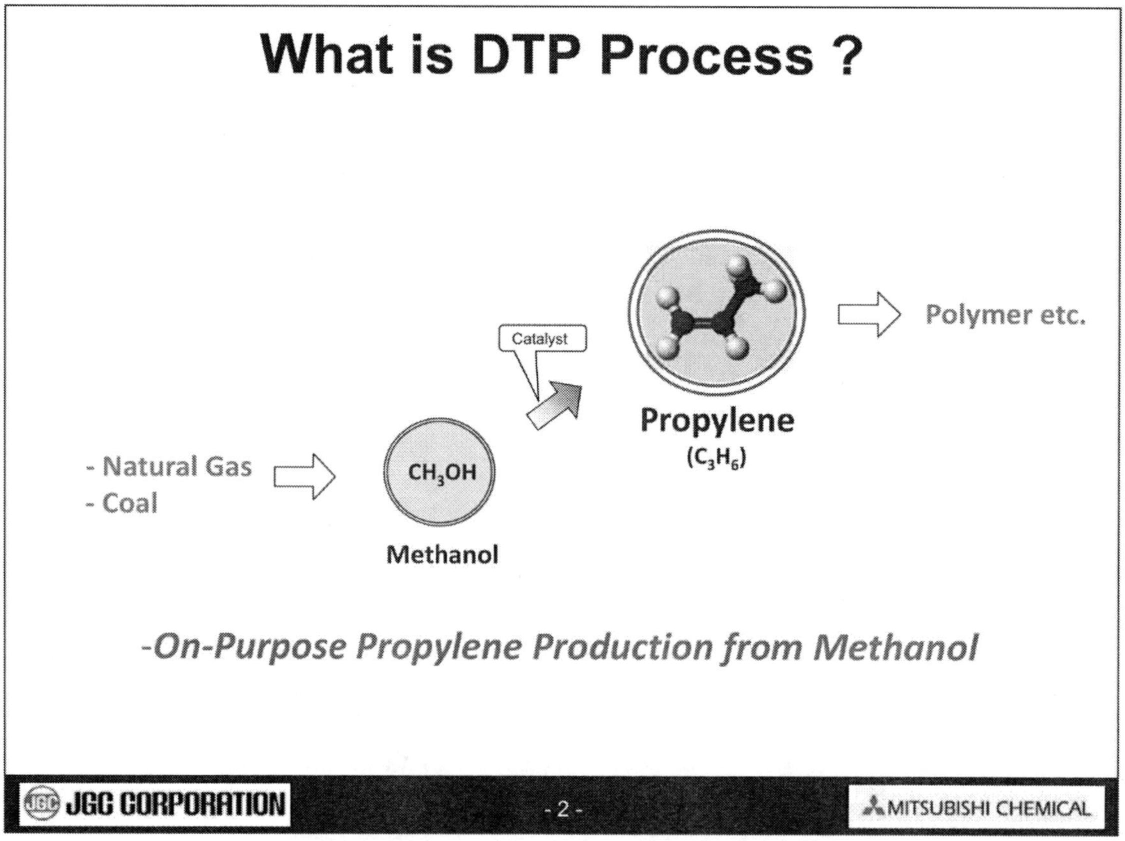
What is DTP Process ?
- Natural Gas
- Coal
CH3OH
Methanol
Catalyst
Propylene
(C3H6)
Polymer etc.
-On-Purpose Propylene Production from Methanol
JGC CORPORATION
- 2 -
MITSUBISHI CHEMICAL

What is DTP Process ?

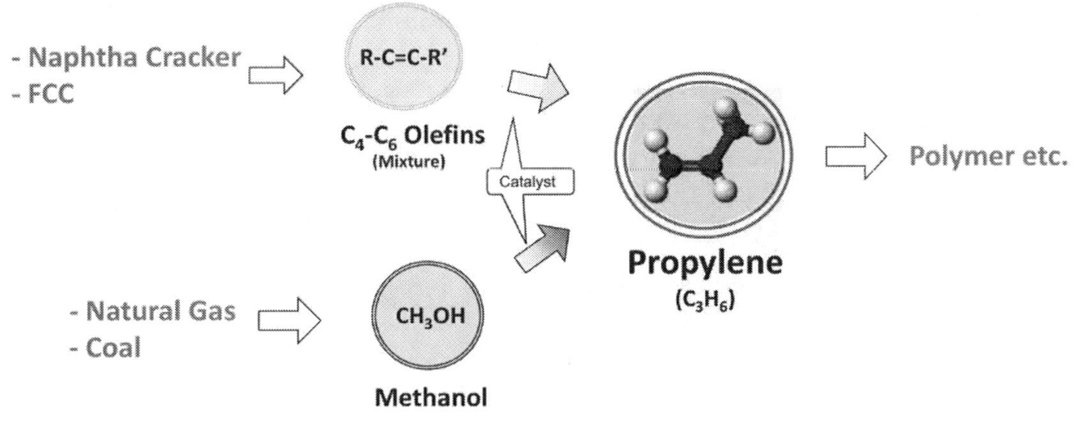

-On-Purpose Propylene Production from Methanol

- Additional Unutilized C_4-C_6 Olefins as Feedstock

Development History and Prospect

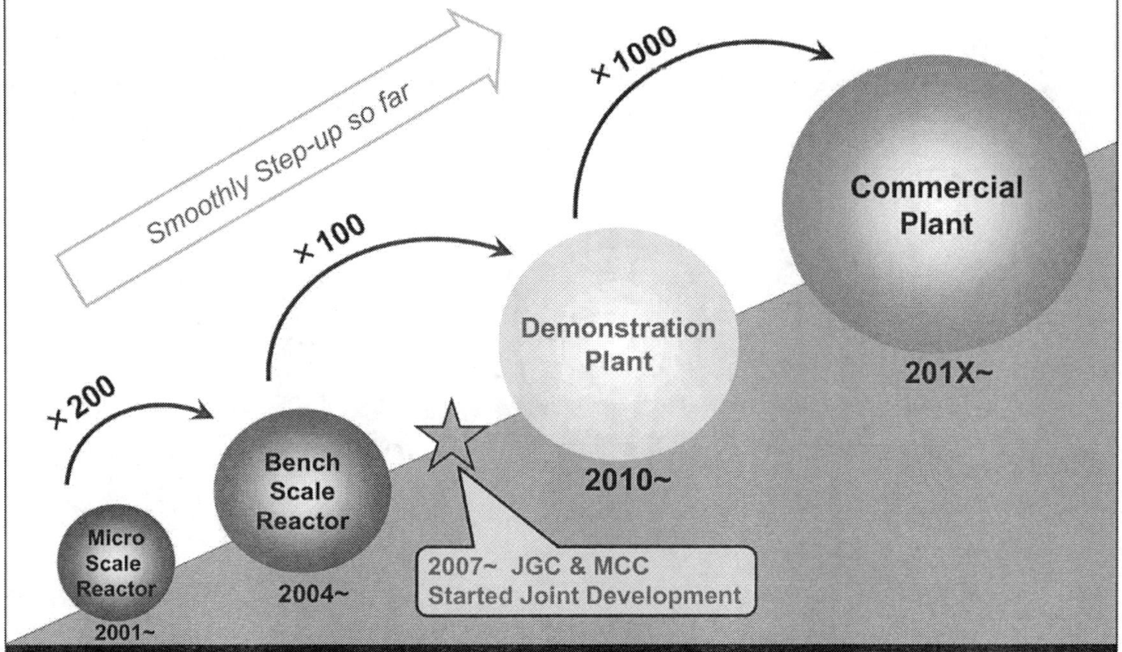

Demonstration Plant Overview

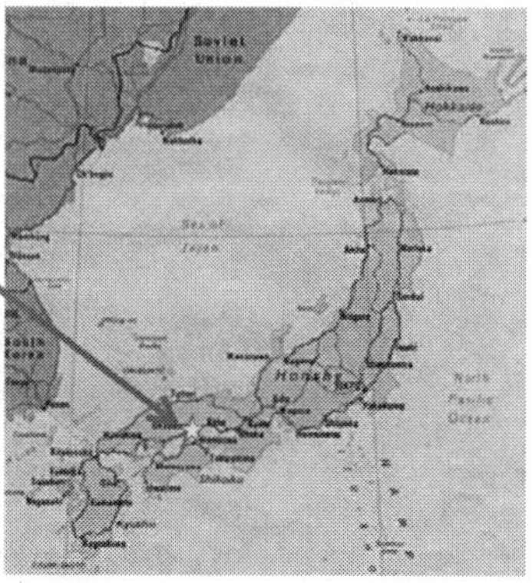

- **Located in Mitsubishi Chemical Mizushima Plant**
- **Successfully Start-up in August 2010**
- **Continuous Operation for Two Years**

Process Flow (Standalone Case)

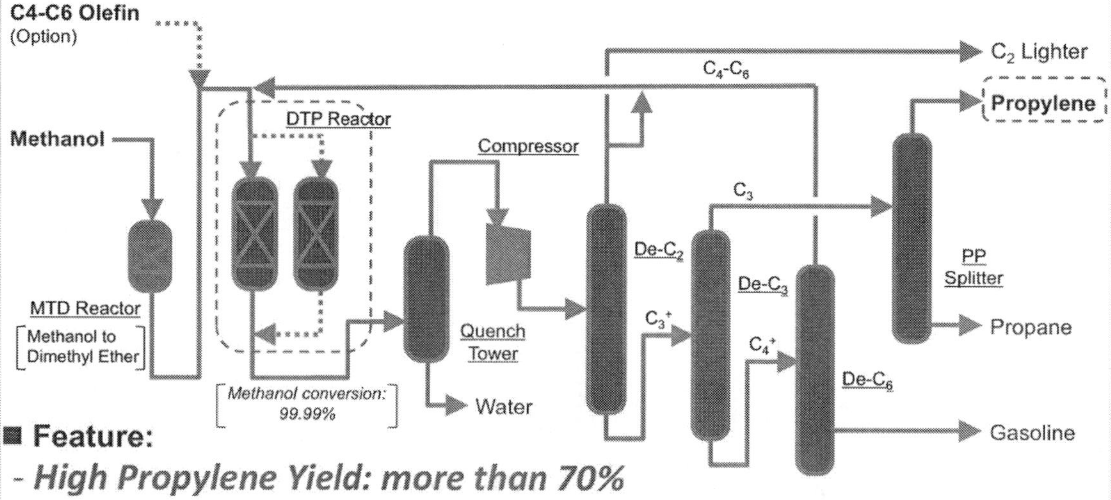

■ **Feature:**

- ***High Propylene Yield: more than 70%***
- ***Applied Existing Proven Technologies***

Process Flow (Integration Case)

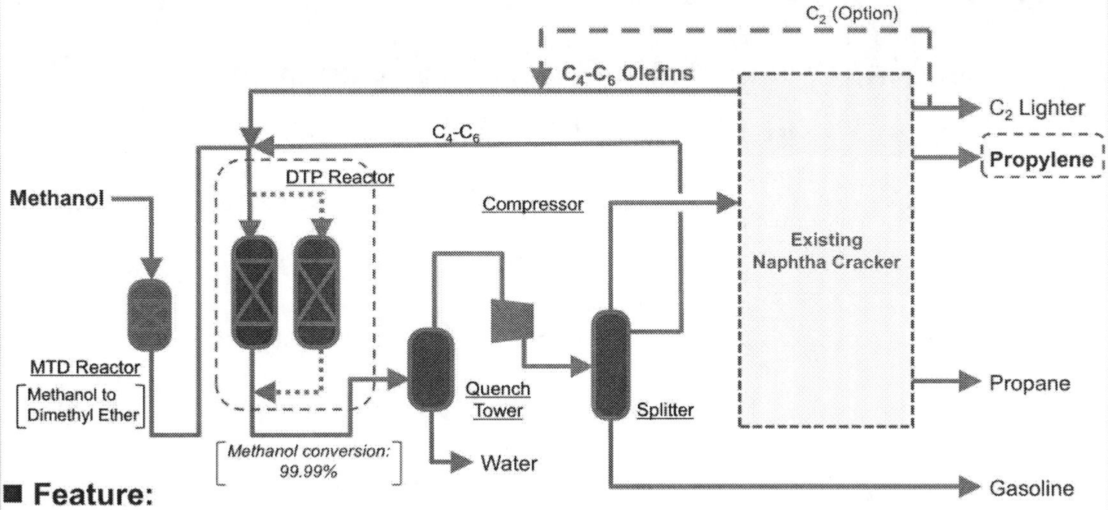

■ **Feature:**

- ***High Propylene Yield: more than 70%***
- ***Applied Existing Proven Technologies***
- ***Low Investment Cost in case of Integration with Naphtha Cracker***

Application

(1) Naphtha Cracker Integration Case (Propylene Capacity: 20KTA – 200KTA)

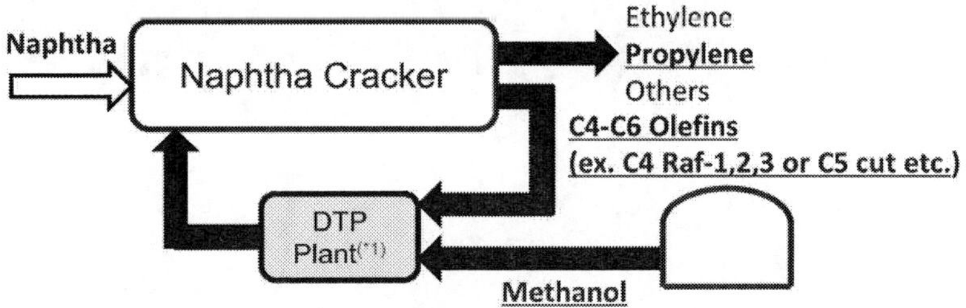

(*1) Utilizing facilities in Naphtha Cracker.

(2) Standalone Case (Propylene Capacity: 250KTA – 500KTA)

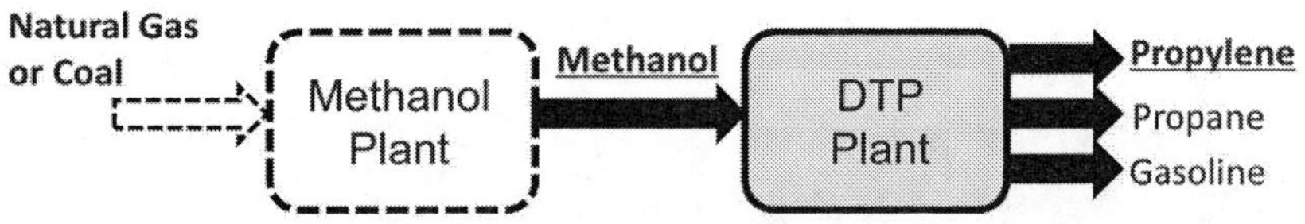

- ***Two Cases available depending on Client requirements !!***

DTP is…

- *High Yield Propylene Production Process*
- *Flexible Feed Methanol to Olefin Process*
- *Process that increase P/E Ratio for Cracker*

Thank you very much !!

Contact:

Mitsuo Morita
Market Development Dept.
Global Marketing Division
2-3-1, Minato Mirai, Nishi-ku,
Yokohama 220-6001, Japan

Phone: 81-45-682-8371
FAX: 81-45-682-8183
E-Mail: morita.mitsuo@jgc.co.jp

JGC CORPORATION - 9 - MITSUBISHI CHEMICAL

352533 Self-Cleaning "Bernoulli" Type Filters Used in Onshore and Offshore Applications

Abstract:

Georg Schuenemann GmbH (SAB-BREMEN) has developed a new Filter Type F-480, dubbed the "Compact Giant" for applications in onshore water intake and cooling water systems, sea water desalination plants, and in offshore projects, such as FPSO's, Drillships, and Production Platforms.

This paper will describe the essential design features of the filters, based on the "Bernoulli principle" which result in a fully automated self-cleaning filter system, and choice of materials in a sea water service with high corrosion rates.

A case study will be presented in which stringent project requirements such as substantial flow rates, adaptibility to space limitations and pipe connections had to be observed. Reliability of the system, extended run times without the need for frequent replacement of filter elements, as well as ease of monitoring and operator involvement have been achieved.

352653 Effects of Support on Sulfur Tolerance and Regeneration of Pt Catalysts Measured By Ethylene Hydrogenation and EXAFS

Abstract:

We investigated the effect of support on sulfur tolerance and regenerability under reducing environments by measuring the rate of ethylene hydrogenation, combined with hydrogen chemisorption, and extended x-ray absorption fine structure (EXAFS). Alumina and titania (P25) supports were used to prepare Pt catalysts at 1 wt.%. The catalysts were tested after treatment in H_2S/H_2 at 250 °C followed by regeneration treatments in H_2 at 250 °C, 350 °C and 450 °C. The sulfided catalysts exhibited a 20-27 times decrease in the rate of ethylene hydrogenation and a 4-6 times drop in the Pt surface area. Regenerations up to 450 °C were unable to remove all the sulfur, as shown by the presence of Pt-S bonds by EXAFS. However, a partial recovery of the hydrogenation rate per mole of Pt was observed on sulfided Pt/Al_2O_3 after reduction at 450 °C, while the induction of strong metal support interactions (SMSI) at reduction temperature above 350 °C was observed on Pt/P25, regardless of the presence of sulfur. For Pt/P25, the reversal of the SMSI state together with sulfur removal by mild oxidation suggests that sequential reduction/oxidation treatments may be more effective in restoring the S-free state of TiO_2-supported catalysts.

352799 Improved Operational Efficiency and Reliability through Insulation Materials Selection

Abstract:

This paper will explore the implications of insulation materials selection on the energy efficiency and reliability for Ethylene production units. Considerations will include insulation properties, structural support, installation methods, installation and maintenance costs.

BOILER MACT COMPLIANCE: MERCURY IN FUEL GAS

PATRICK LAINE, WES YONGE, & VIC VICKERY
PEI, HOUSTON, TX 77043

Patrick Laine, Ph.D.
Chief Scientist
Houston, TX
850-866-6686
plaine@pei-tx.com

Wes Yonge, PE
Chief Engineer
Houston, TX
713-806-8396
wyonge@pei-tx.com

James 'Vic' Vickery
Chief Technology Officer
Houston, TX
281-536-0899
vvickery@pei-tx.com

1 Introduction & Objectives

For most ethylene producers the primary objective of this sampling and analysis effort will be to verify and "establish a mercury fuel specification through which gas-fired boilers that use a fuel other than natural gas or refinery gas may be considered Gas 1 unit" as identified in 40 CFR Part 63.7521.(f).

Measurement and Monitoring Requirements: Decision Rules

❑If Hg concentration ≤ 50% of Gas 1 specification* then no further testing required

❑If Hg concentration > 50% and ≤ 75% of the specification* then semi-annual sampling required

❑If Hg concentration > 75% of the specification* then monthly sampling required

*Specification is **40 μg/Sm³**

The goal will be to execute a sampling and analysis campaign that will produce compliance quality data. The key to producing and validating compliance quality data is a well-designed and executed QA/QC protocol. At a minimum, a performance based QA/QC protocol is recommended that includes the following components:

- Sorbent Trap Breakthrough Bed:
  - ❑ Sorbent traps consisting of a minimum of two sorbent beds in series so that the trailing bed can be analyzed to verify that no breakthrough occurred from the leading bed.
- Duplicate Samples:
  - ❑ The sampling system should be configured with two trains to accommodate the collection of simultaneous duplicate sorbent trap samples, in parallel, providing a means of verifying precision in the sample collection process.
- NIST (National Institute of Standards and Technology) Traceable Spike:
  - ❑ Sorbent traps with a third bed (downstream) that is spiked with a NIST traceable spiked mass of mercury serves as a matrix spike and provides a means of verifying that components in the matrix (the fuel gas) are not affecting the trapping efficiency of the sorbent trap.

2 Methods

Sorbent Trap Mercury Sampling Methods Comparison

Method	Description	Method Type: S = Sampling A = Analysis	Evaluation	Sample Collection Phase QA/QC
ISO 6978	Hg in Natural Gas by Thermal Desorption (CVAFS)	S & A	Method calls for low sample volumes can lead to non representative results.	No
EPA Method 30B	Determination of Gaseous Hg Emissions from Coal Fired Combustion Sources using Carbon Sorbent Traps	S	Requires modification (Modified M30B)	Yes
ASTM D6350	Hg S&A in Natural Gas (AFS)	S & A	Method calls for short term samples (i.e., low sample volume) which leads to poor accuracy.	No
ASTM D5954	Hg S&A in Natural Gas (AAS)	S & A	AAS is less sensitive than AFS.	No
EPA Method 1631	Measurement of Hg in Water	A	No	Yes

3 EPA Method 30B QA/QC Performance Criteria

QA/QC Test or Specification	Acceptance Criteria	Frequency
Analyzer calibration	Correlation Coefficient: 0.999	Daily
Transfer efficiency of analyzer	%Transfer Efficiency: 90%–110%	Once per sorbent trap
Initial calibration blank	< 0.5 ng	Once before each calibration
Sorbent Trap Section B breakthrough	≤ 10% of Section A Hg mass, for Hg concentrations > 1μg/dscm ≤ 20% of Section A Hg mass, for Hg concentrations ≤ 1μg/dscm	At each sample point
Spike Recovery Test	Average recovery between 75% and 125 % for Hg_0	Once per sample point
Paired Sorbent Traps Agreement	≤ 10% RD for Hg concentration > 1μg/dscm ≤ 20% RD or 0.2 μg/dscm absolute difference for Hg concentrations ≤ 1μg/dscm	At each sample point

4 Equipment

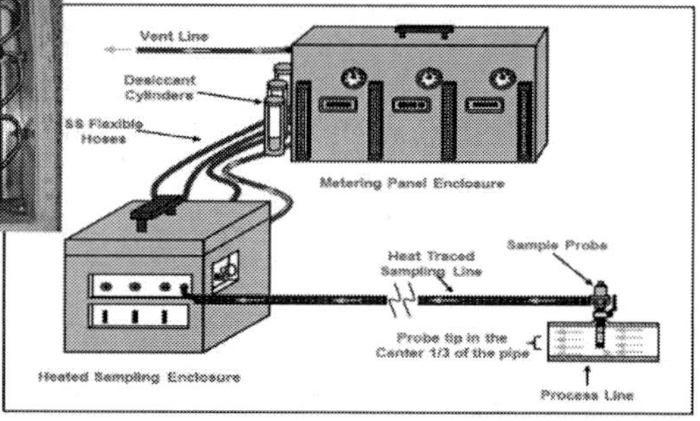

Flow

QA/QC Acceptance Criteria:	Primary Bed	Breakthrough Bed	Spiked Bed
	99% of Total Collected Hg	≤ 10% of Primary Mass	Recovery between 75% and 125%

Gold Trap:

5 Conclusions/PEI Recomendations

❑ There are five methods accepted by the Boiler MACT rule. PEI has extensive experience and expertise employing **all five** of these methods.

❑ Operators can achieve compliance without submitting a Fuel Analysis Plan by using ASTM D6350, however, we have noticed variable results with this method.

❑ The **PEI recommendation** for obtaining the most accurate compliance quality data is to use Modified EPA Method 30B in conjunction with EPA 1631 where traps should be analyzed in a certified ISO 17025 laboratory. The recommended pathway includes a rigorous, performance based QA/QC protocol that validates accuracy and precision of all sample collection procedures.

351376 Young Professional Simulation Tutorial

Abstract:

18C00 Young Professional Simulation Tutorial

Authors: Naomi Hua and Mike Donahue, Invensys

Process simulation is a complex skill to master. Even with years of experience, a typical user is constantly facing new problems that need to be addressed. There are many issues that can lead to unexpected and unnecessary errors with the use of process simulation tools. Regardless of what simulation program you are using, these issues are common from one program to another. In this presentation, targeted for the young professional, we will discuss the "Most Common Pitfalls in Process Simulation". For each topic, we will discuss various techniques and methods these troublesome problems and prevent their occurrence.

Topics to be discussed include:

1. Erroneous data entry
2. Simulation defaults
3. Bad plant data
4. Thermodynamic method selection

5. Simulation tolerances and convergence

6. Recycle identification and handling

7. Model complexity

8. Difficult or impossible specifications

9. Assay pseudo-component development

10. Assay property blending

1. Erroneous data entry is the most common mistake made when building a process simulation model. This presentation will cover methods that can help ensure proper data entry in your process simulation model, such as checking the data you entered, particularly the units of measure, using "alternate forms" of data input such as copy/paste and data import, and reviewing all warning and error messages.

2. Defaults are methods, values, and conditions that are automatically selected for you by the program when you begin a new simulation model. These defaults will affect the results calculated by your model and every simulation program has them. Most defaults within simulation programs fall under categories such as convergence criteria, thermodynamic methods, tolerance and units of measure. The problem with defaults is unintended results from not fully understanding the effects a particular default has on the model. Take the time to understand the defaults of the program and how these defaults will affect the results. If you do not agree with the defaults selected by the program, most simulation programs will allow the user to change the default.

3. Plant data changes over time. The feed information and process conditions of the unit operations may be different. Not only does plant data change over time, but measurement information is often very inaccurate. This could be due to faulty or poorly calibrated measurement equipment, leaks in lines, sensor failures, and other issues. Because of these changes and inaccuracies over time, you always have to validate the information that is entered in your model to ensure that you are obtaining the proper results. To do so, you can utilize programs that can help with data validation and reconciliation like SimSci's ROMeo online optimization program. With it, you can connect your model to your data historian to constantly import current data measurements into your model and then reconcile these measurements to determine when problems arise within your plant and where those problems are occurring. You will also need to use your engineering judgment to determine if the data is valid. More details will be provided during the presentation.

4. The proper selection of thermodynamic methods for your simulation is the most important decision that you make when modeling a process. The method you select could have a drastic effect upon your results. Often, any number of thermodynamic methods is suitable for a given application. It is up to the user to determine the best method to use based upon knowledge, personal experience, and best judgment. Some important factors that need to be considered are the components being used and the operating conditions.

Oftentimes, the selected thermodynamic method still needs to have additional inputs such as binary interaction parameters to simulate the process accurately.

5. Many of the calculations in process simulation are iterative. When these calculations iterate, the program compares the results calculated from one iteration to the results calculated in the next iteration in order to determine

whether it is converged to a solution or needs to continue calculating. This method of comparing results is where tolerances come into play and simulation programs use tolerances to determine when a solution is reached. One of the major mistakes when modeling a process is not understanding or using the correct tolerance values. A looser tolerance may make the file easier and quicker to converge, but it also reduces the accuracy of the model. A tighter tolerance may be more accurate, but might make it impossible for the model to solve or increase the convergence time exponentially. A common practice is that when you have reached a converged solution with your model, tighten the tolerances within the model and solve it again. If you notice that your answers have changed significantly, you need to continue tightening the tolerance values until your results change little from one iteration to another. You will then know that you have reached an accurate solution for your model.

6. Recycles in process simulation occur when you return to a previously calculated unit operation to recalculate the unit due to a change made elsewhere in the model. They are very common and can often be difficult to converge. Some suggestions on troubleshooting recycle convergence include initially solving your model without the recycles to see how it behaves, providing initial estimates for recycle streams, using tighter tolerances on unit operations within recycle loops than the total recycle tolerances themselves, and ensuring that all components have a way in and out of a recycle by performing mass balances and using purge or makeup streams where necessary.

7. The general rule of thumb when modeling is to start simple and add complexity as you go. Each time you add a few streams or unit operations converge and save the model before moving on. Make sure you justify any additions that you make to the model. You can most likely model your system without including every unit from your PFD in the simulation. When modeling, think about the goal of your simulation and not about literally translating every real-world piece of equipment into the simulation model. Make the judgment call on what is most important and what the purpose is of every unit you add to your simulation. For example, if your real-life process shows a heat exchanger, a pump and a splitter, but all that is necessary is that the temperature, pressure, and separation are achieved properly, consider using a flash drum to accomplish all three.

8. Specifications within process simulation are everywhere and every stream and unit operation you place in your model contains specifications. In its simplest form, a specification is a constraint that you place on the model that the model needs to meet in order to successfully converge. For streams, these are usually temperatures, pressures, flowrates and compositions. For unit operations, they range from process conditions, like temperatures and pressures, to product conditions, like an overhead composition in a column. There are some pitfalls, however, that you can run into with specifications such as azeotropes in columns, high purity specifications, and the possibility of multiple solutions.

9. Petroleum assays are sets of distillation data that are used to characterize all of the components that reside in a given crude sample. When you enter distillation data into a process simulator, the program converts it into components that it can use when modeling your process. The components that are generated are called pseudo-components, and each pseudo-component represents all of the unknown components that would be within a given boiling point range in the crude. Some of the things you will want to ensure when properly developing your assay components are did you use the correct distillation curve basis, did you enter the data correctly, did you use enough cuts and are you using the correct characterization methods .

10. When you have more than 1 assay in a simulation model, you need to be concerned with assay blending and how it will affect the results of your file. Assay blending is the mixing of 2 or more groups of components (i.e. "C sets of cuts/pseudo-components) in order to calculate averaged component properties for cuts that are in the same boiling point range as one another. Once these new averaged component properties are determined, the program will use these averages from this point on in the simulation model any time it needs to access component

properties for a given pseudo-component. All component properties are blended, including normal boiling points, specific gravity, molecular weight and refinery inspection properties, and usually the blended component properties are calculated using a weighted average

These topics are based on experience and feedback from end-users to technical support representatives for process simulators. They often represent the most commonly asked questions or areas that most users find to be problematic. While the specific examples highlighted in each topic come from SimSci software, the intent of this tutorial presentation is to be broad-based and is particularly applicable to young professionals seeking to develop their experience with simulation.

Polymeric Sulfonic Acid Composites for Fuel Cell Applications: Effects of Casting Solvent and Sonication Time on Membrane Performance

Jimoh K. Adewole[1], Abdullah S. Sultan[1,], Amir Al-Ahmed[2], and S. M. Javid Zaidi[3]*

[1]Center for Petroleum & Minerals, King Fahd University of Petroleum & Minerals, Dhahran 31261, Saudi Arabia

[2]Center of Research Excellence in Renewable Energy, King Fahd University of Petroleum & Minerals, Dhahran 31261, Saudi Arabia

[3]School of Chemical Engineering, The University of Queensland, Australia

Abstract

Polymer blends comprising sulfonated poly (ether ether ketone) (SPEEK) and 3wt% (2-acrylamido-2-methyl-1-propanesulfonic acid) (PAMPS) were prepared by solution casting method. Two solvents (N,N-Dimethylacetamide (DMAc) and N-Methyl-2-pyrrolidone (NMP)) were uesd separately. Residual solvents in the membrane samples after drying were quantified using thermo gravimetric analysis. Effects of periods of sonication (0 to 90 minutes) and types of solvent on the membrane performance were evaluated interms of morphology and water uptake. An attempt was made to correlate membrane performance with the time of sonication in order to select the most suitable solvent and membrane casting conditions for subsequent membrane preparations. For both solvents, the water uptake was observed to first decrease at 10 minutes of sonication and then increased as the sonication time was increased. The minimum water uptake recorded for DMAc samples was 19.23% at 20 minutes sonication time while that of NMP sample was 24.24% at 10 minutes. However membrane morphology was found to deteriorate with increase in sonication time. No correlation was obtained between the membrane properties and the sonication time. In general, membrane performance was better for membranes samples using DMAc than NMP solvent. Therefore, DMAc was selected as the casting solvent for membrane preparation on which further studies will be carried out.

Keywords: morphology, water uptake, residual solvent, thermogravimetric analysis

Introduction

Water soluble polymers have received academic and industrial interests due to their potential applications in fuel cells and other electrochemical devices (Ye et al. 2012). One of the water soluble polymers that have attracted publication interest is poly (2-acrylamido-2-methyl-1-propanesulfonic acid) (PAMPS) (Florjańczyk et al. 2003). Generally, some of the major performance criteria for fuel cell materials include proton conductivity, water and methanol uptake, good mechanical properties, chemical stability, thermal stability and methanol permeability (Gil et al. 2004; Swier et al. 2005; Ye et al. 2012). PAMPS was reported to posses higher proton conductivity than partially hydrated Nafion. In addition, it has been found to be more tolerant to water fluctuation in high temperature operations. Some of the major drawbacks in the use of PAMPS are its poor mechanical properties, thermal stability, water and methanol uptake and solubility in many solvents (Florjańczyk et al. 2003). One of the approaches that have been proposed to solve these problems is development of polymer composites.

This research paper investigates the synergistic effect of combining water soluble PAMPS with sulfonated poly (ether ether ketone) (SPEEK) which is non - water soluble polymer to prepare composite membranes for application in fuel cells. Specifically, preliminary results on the performnance of the composite were presented. Performance were evaluated based on the composite morphology and water uptake. Effects of sonication time and types of casting solvent on membrane properties were also evaluated.

Material and Method

Material

Sulfonated poly (ether ether ketone) (SPEEK) purcahsed from FuMA-Tech, Germany, and poly (2-acrylamido-2-methyl-1-propanesulfonic acid) (PAMPS) purchased from Aldrich Chemical Co., are the two major polymers used in this research. The SPEEK has a DS of 0.533 and solubility of 99% in DMAc. SPEEK is derived by sulfonation of Poly (ether ether ketone) (PEEK). PEEK is a commercially successful polymer due to its excellent thermal stability and solvent resistance (Mohr et al. 1991). For sometimes, further application of PEEK have been stalled due this solubility issues. By sulfonation, PEEK can be made soluble in most commercial solvents. Importantly, the solubility of SPEEK also depends on its degree of sulfonation (DS). Since, this work involves the use of two polymers, a solvent which will disolve both the PAMPS and SPEEK must be selected. Literature review showed that PAMPS dissolves in N,N-Dimethylformamide (DMF), N,N-Dimethylacetamide (DMAc), propylene carbonate (PC), and water (Florjańczyk et al. 2003). Initial solubility studies that was carried out in our laboratory was done for N,N-Dimethylacetamide (DMAc) and N-Methyl-2-pyrrolidone (NMP). The choice of NMP was due its unique solvent properties and its structtural distiction from DMF and DMAc (Figure 1). Moreover, no literature was found where NMP was used as casting solvent for PAMPS. Thus, DMAc and NMP were used as solvents in this work. All materials were used as received from the manufacturers.

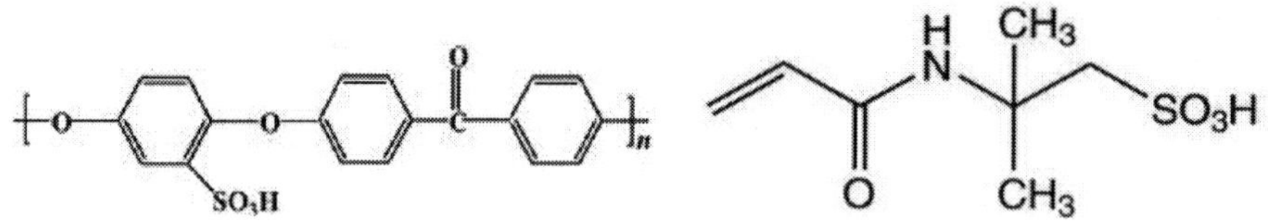

Figure 1 Structure of SPEEK and PAMPS

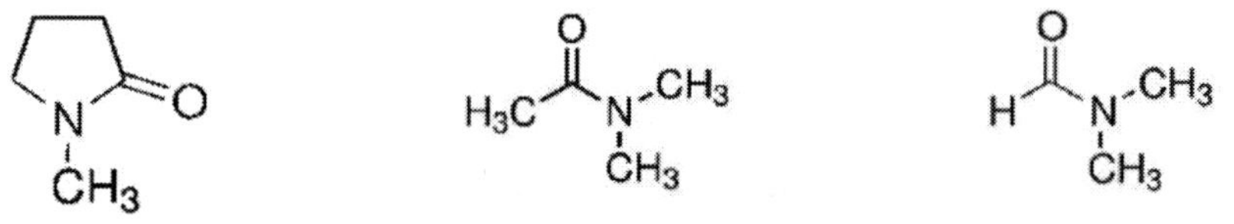

N-Methyl-2-pyrrolidone N,N-Dimethylacetamide N,N-Dimethylformamide

Figure 2 Structure of Solvents

Experimental Procedure

Membrane Preparation

SPEEK was first dissolved in solvents (DMAc and NMP) and stirred for 2hours at 70°C. PAMPS was then added and the blend was stirred for 24hours. The solution was sonicated for various periods of time ranging from 10 to 90minutes using an Ultrasonic bath. In untralsonic bath, the final temperature achieved for a given time of sonication depends on the tempearutre of the laboratory (summer or winter). It is therefoe genrally advisable that maximum temperature of the bath is determined and maintains (as the so - called equilibrium tempearture), when operating continuously under ambient conditions (Santos et al. 2009). During the sonication, equilibrium temperature was maitained by simply filling the bath with water that was preheated to sample temperature.

Membrane Characterization

Water uptake and morphological tests were performed. Water uptake was carried out by first vacuum drying all membrane samples at 100°C until constant weights were obtained. Dried samples were then immersed in deionized water for 24h at room temperature (25°C). Samples were taken out of the water after 24h, wipped to remove surface water and then weighed immediately. Water uptake was calculated using Eq. 1.

$$water - uptake = \frac{W_{wet} - W_{dry}}{W_{dry}} \quad \text{Eq. 1}$$

Surface morphology of the membrane was done using scanning electron microscopy (SEM). Thermogravimetric analysis (TGA) was done from room temperature to 350°C using TA Q600 analyzer at heating rate of 20°C/min under nitrogen atmosphere with flow rate of 20mL/min using ~ 8 -10mg membranes samples.

Results and Discussion

Membrane Morphology by Scanning Electron Microspcopy (SEM)

Results of surface morphological analysis performed on membrane were shown in Figure 3 and 4. The overall pictures of this figure depicts a dense membrane with some surface defects observed for most of the sonicated samples. For example, Figure 3A showed some kind of layered arrangement with no visible pores at the surface. Also, the PAMPS seems to be well distributed throughout the SPEEK. The micrograph in Figure 3B can be observed to be rough and possesses some uneven ramdom distribution of PAMPS agglomerates indicating that the distribution of PAMPS have been disturbed by the 10minutes sonication. An increase in sonication time to 20 minutes was observed to improve the distribution of PAMPS. This was however accompanied by some cracks believed to be due to the slight increase in the viscousity of the blend solution. The vicousity increase was brought about by the excape of some fraction of the solvent during sonication. As the sonication time was increased, PAMPS was observed to become more evenly distributed and the cracks become more visible and larger in size (Figure 3D to F). SEM images of samples at sonication time of 10 and 30 minutes were reproduced in higher magnification of 10μm (Figure 4) in order to further revealed the effect of sonication on the morphology of the somposites. Similar morphological behaviour has been reported in literature, though for inorganic/organic clusters of nanoclay/epoxy composites membranes (Lam et al. 2005).

Water Uptake

Water uptake have been reported to be affected by the degree of sulfonation, membrane pretreatment, hydration, ambient relative humidity and water temperature (Xing et al. 2004; Wu et al. 2006). Figure 5 showed the effect of sonication time and solvent type on water uptake that was measured at room temperature (25°C). For both solvents, water uptake is same at zero sonication time. There was however a further decrease in the water uptake for NMP at 20 minutes sonication time before it beging to increase. Thus, the best sonication time for both solvent lied between 10 and 20 minutes with respect to water uptake. Interestingly, this result correspond to that reported by Lam (2005) on the effect of ultrasonication on microhardness of composites.

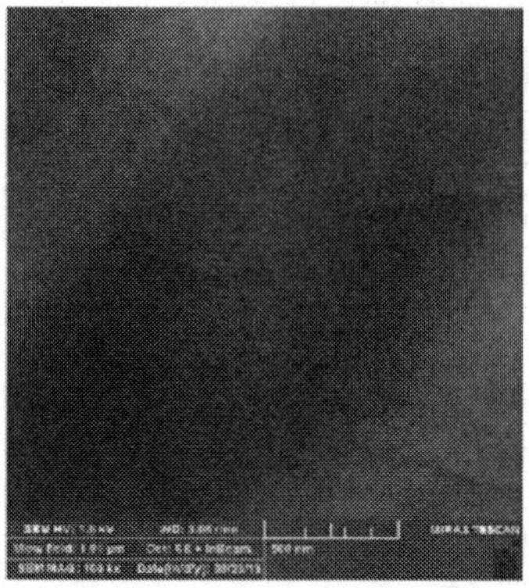

A) No sonication

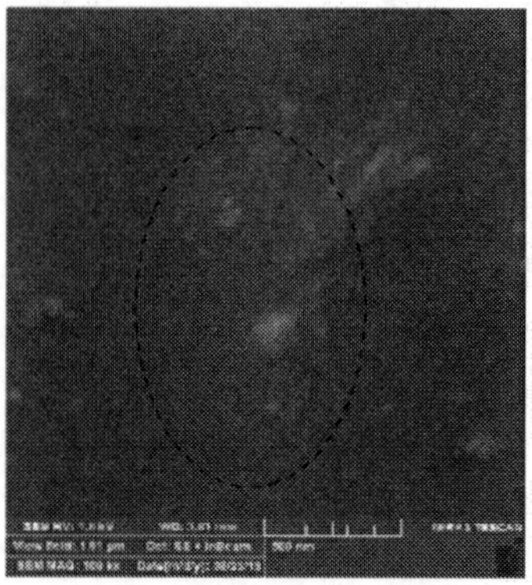

B) 10 minutes sonication

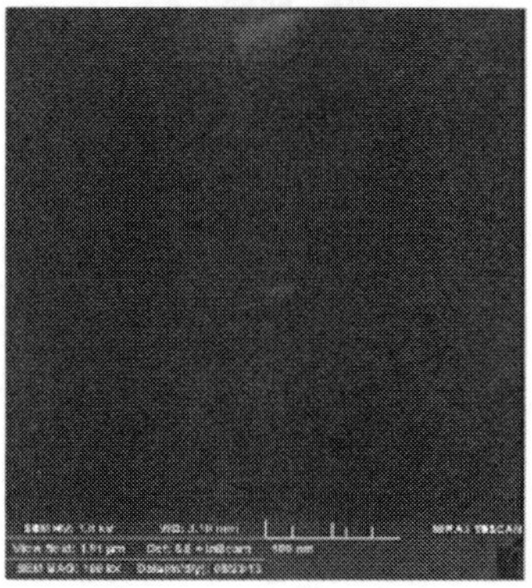

C) 20 minutes sonication

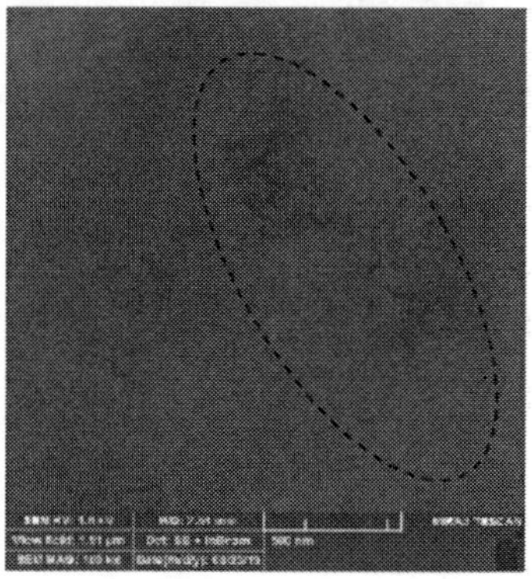

D) 30 minutes sonication

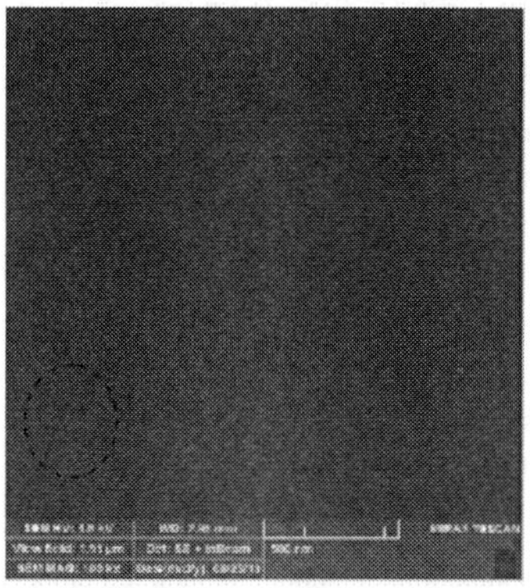

E) 60 minutes sonication

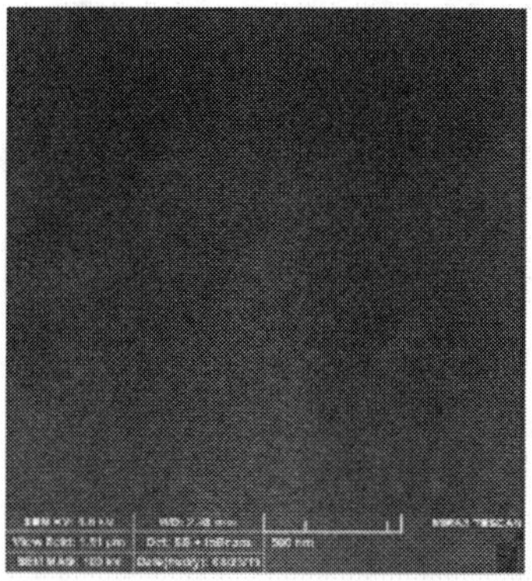

F) 90 minutes sonication

Figure 3 SEM micrograph of SPEEK-PAMS composite with various times of sonication with magnification of 500nm

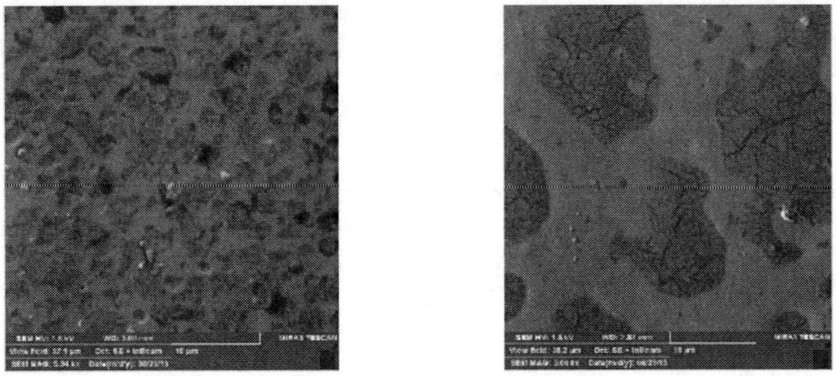

A) 10 minutes sonication B) 30 minutes sonication

Figure 4 SEM micrograph of SPEEK-PAMS composite with various times of sonication with magnification of 10μm

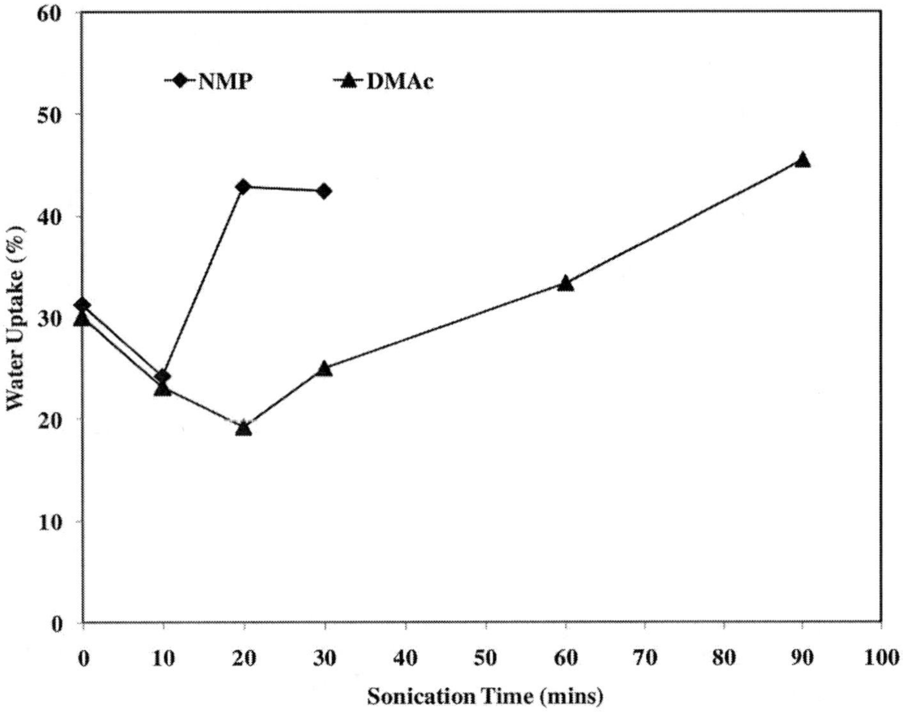

Figure 5 Water uptake of SPEEK-PAMS (3wt%) composite with variuos sonication times

Thermogravimetric analysis (TGA)

The difference in membrane performance resulting from the use different casting solvents have been attributed to the presence of residual solvents in the membrane samples after evaporation (Joly et al.

1999). TGA was thus used to quantify the residual solvent and evaluate its effect on the membrane performance as has been previously demonstrated in lieterature (Recio et al. 2007; Iqbal et al. 2008). Figure 6 shows the TGA curve for samples prepared from NMP and DMAc with different post vacuum heat treatment conditions. From this figure, weight loss can be attributed to evaporation of variety of substances, including solvents, from the samples (Rich et al. 1997). Table 2 contains the solvent and its boiling points, and weigth loss at the boiling point range that were extracted from TGA results. For sample heat treated for 24hrs, the overall percentage weight loss up to 204°C was about 8wt% of membrane sample for NMP while it is about 6.3wt% for DMAc. These will definitely include compnenents such as water, solvents and other unidentified substances. To quantify residual solvents, weight loss was estimated between 192 -214°C for SPEEK-PAMPS-NMP and between 154 and 176 for SPEEK-PAMPS-DMAc. The results obtained showed that the amount of the residual solvent in NMP is larger than that of DMAc by 1.25wt% (Table 1). To further investigate if this difference is responsible for higher value of water uptake for the NMP sample, membrane samples made from both solvents were further heat treated under vacuum for 72hrs. Residual solvent after this treatment was slightly higher in DMAc than NMP. Despite this, the value of water uptake of the NMP membrane is still higher than that of DMAc by approximately the same value as before the heat treatment (1.71wt%).

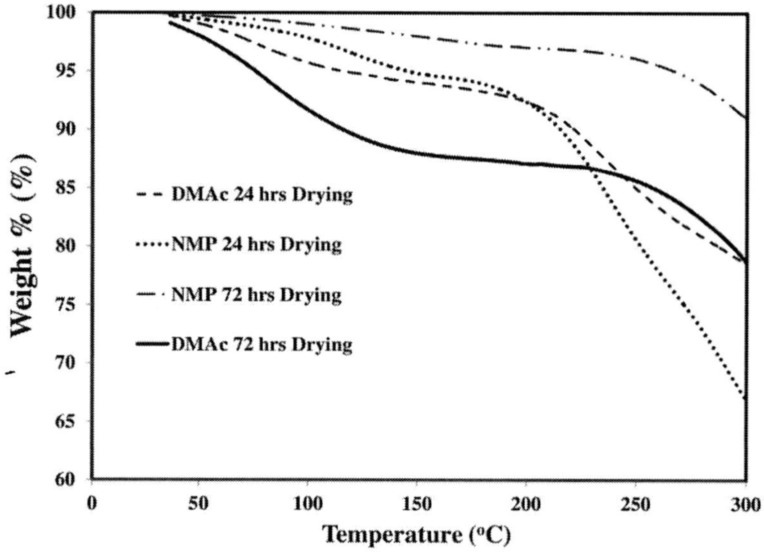

Figure 6 TGA of SPEEK-PAMPS-NMP and SPEEK-PAMPS-DMAc

Table 1 Residual solvent in SPEEK-PAMPS composites membrane samples

Membrane Sample	Solvent	B. Pt. (°C)*	Residual Solvent (wt%)	Water Uptake
DMAc 24 hrs Drying	DMAc	164.5 – 166	0.57	30.00
NMP 24 hrs Drying	NMP	202 – 204	2.85	31.25
DMAc 72 hrs Drying	DMAc	164.5 – 166	0.367	21.21
NMP 72 hrs Drying	NMP	202 – 204	0.217	22.92

*Boiling Points of Solvents

On a general note, results from Figire 1 – 5 showed that the best perfomance can be obtained for variety of sonication time depending the property of preference. Also, performance of DMAc solvent was observed to be better than that of NMP on the basis of water uptake. For this reason, DMAc was selected as the casting solvent for membrane preparation on which further studies will be carried out. Water uptake of DMAc membrane, prepared without sonication, was comapred with those reported in previuos literature. Table 2 illustrates the comparison between SPEEK-PAMPS and other relevant membranes previuosly reported in literature. In addition, this table serve to provide insight into the possible potentials of the composites. From this table, it was observed that the water uptake for SPEEK-PAMPS is within the range of values of other membranes on the table. Its water uptake is less than that of Nafion 112 which is a commercial membrane. The differences observed between the water uptake values of SPEEK – PAMPS and other membranes from the table can be explained based on three important factors - SO_3H group, the degree of sulfonation and experimental temperature (Gil et al. 2004; Shan et al. 2006; Wu et al. 2006). Water uptake of SPEEK – PAMPS is higher due to the presence of SO_3H group in the PAMPS molecules (Figure 2). SO_3H group have been reported to enhance membrane acidity and hydrophilic properties (Shan et al. 2006). Infact, the higher the SO_3H group the larger the water uptake. The degree of sulfonation has been reported to affect the water uptake of membranes. The higher the degree of sulfonation, the larger the value of water uptake (Gil et al. 2004). Thus, water uptake of SPEEK – PAMPS is lower than SPEEK -3 due their different degree of sulfonation, 53.3 and 72% respectively and the experimental temperature 24 and 60°C, respectively. The experimental temperature of SPEEK – 2 is close to that of the composite. However, the results cannot be compared in this case since the DS of SPEEK – 2 was not known. Above explanation is also true for the values of water uptake for SPEEK – 1.

Table 2 Comparison of water uptake values obtained with those from Literature (Xing et al. 2005)

Membrane	Water Uptake (wt%)	Experiment Temperature (°C)	**D S (%)	Reference
Nafion 112	22.85	N/A	-	(Xing et al. 2005)
SPEEK – 1*	16.00	30	45	(Ren et al. 2005)
SPEEK – 2*	44.90	25	N/A	(Zhong et al. 2008)
SPEEK – 3*	85.00	60	72	(Wu et al. 2006)
SPEEK– 4*	15.00	N/A	N/A	(Handayani and Dewi 2011)
SPEEK/PAMPS (3wt%)	21.21			This work

*The numbering have been included for explanation purposes.
**DS means Degree of Sulfonation

Conclusion and Recommendations

Blend of SPEEK and PAMPS was prepared by solvent casting method and its performance was evaluated with respect to its morphology and water uptake. Effect of sonication time (10 – 90minutes) and solvent types (DMAc and NMP) were also investigated. Performance of membrane prepared using DMAc was observed to be better than that of NMP. There was no correlation between the sonication time and membrane properties, but it was observed that the sonication deteriorate the membrane morphology. Membrane prepared without sonication shows better surface morphology, whereas the best water uptake was observed between 10 and 20 minutes. Thus, other properties need to be investigated in order decide the best sonication time. Therefore, further investigation such as thermal stability, methanol permeability, ion exchange capacity and gas (O_2 and H_2) permeability of the membrane are recommended for a complete evaluation of the membrane performance for direct methanol fuel cell (DMFC) application. All these investigations are in progress in our laboratory.

Acknowledgement

Authors would like to thank the National Science, Technology and Innovation Program (NSTIP) of KACST for funding this research under project number 10-ENE1374-04. Also, authors extend their appreciation to Center of Petroleum & Minearls, Center of Excellence in Renewable Energy, Mr. Dan Jerry D.Cortes of the Nanotechnology Laboratory, Chemical Engineering Department, King Fahd University of Petroeleum & Minerals, Saudi Arabis for their supports.

REFERENCE

Florjańczyk, Z., et al. (2003). "Proton conducting electrolytes based on poly(2-acrylamido-2-methyl-1-propanesulfonic acid)." Electrochimica Acta **48**(14–16): 2201-2206.

Gil, M., et al. (2004). "Direct synthesis of sulfonated aromatic poly(ether ether ketone) proton exchange membranes for fuel cell applications." Journal of Membrane Science **234**(1–2): 75-81.

Handayani, S. and E. L. Dewi (2011). "Influence of Silica/Sulfonated Polyether-Ether Ketone as Polymer Electrolyte Membrane for Hydrogen Fueled Proton Exchange Membrane Fuel Cells." Internat. J. of Sci. and Eng. **2**(2): 26-30.

Iqbal, M., et al. (2008). "Solvent effect on morphology and CO2/CH4 separation performance of asymmetric polycarbonate membranes." J. Membr. Sci. **318**(1–2): 167-175.

Joly, C., et al. (1999). "Residual solvent effect on the permeation properties of fluorinated polyimide films." Separation and Purification Technology **16**(1): 47-54.

Lam, C.-k., et al. (2005). "Effect of ultrasound sonication in nanoclay clusters of nanoclay/epoxy composites." Materials Letters **59**(11): 1369-1372.

Mohr, J. M., et al. (1991). "Gas transport properties of a series of poly(ether ketone) polymers." Polymer **32, 13**: 2387-2394

Recio, R., et al. (2007). "Gas separation of 6FDA–6FpDA membranes: Effect of the solvent on polymer surfaces and permselectivity." J. Membr. Sci. **293**(1–2): 22-28.

Ren, S., et al. (2005). "Surface modification of sulfonated poly(ether ether ketone) membranes using Nafion solution for direct methanol fuel cells." Journal of Membrane Science **247**(1–2): 59-63.

Rich, D. C., et al. (1997). "Effect of cure conditions on Probimide 32 polyamide-imide." Journal of Applied Polymer Science **63**(9): 1113-1126.

Santos, H. M., et al. (2009). The Power of Ultrasound. Ultrasound in Chemistry: Analytical Applications. J.-L. Capelo-Martinez. Weinheim, Wiley-VCH.

Shan, J., et al. (2006). "Sulfonated polyether ether ketone (PEEK-WC)/phosphotungstic acid composite: Preparation and characterization of the fuel cell membranes." Pure Appl. Chem **78**(9): 1781 - 1791.

Swier, S., et al. (2005). "Sulfonated poly(ether ketone ketone) ionomers as proton exchange membranes." Polymer Engineering & Science **45**(8): 1081-1091.

Wu, H.-L., et al. (2006). "Preparation and characterization of poly(ether sulfone)/sulfonated poly(ether ether ketone) blend membranes." European Polymer Journal **42**(7): 1688-1695.

Xing, D. M., et al. (2005). "Characterization of Sulfonated Poly(Ether Ether Ketone)/Polytetrafluoroethylene Composite Membranes for Fuel Cell Applications." Fuel Cells **5**(3): 406-411.

Xing, P., et al. (2004). "Synthesis and characterization of sulfonated poly(ether ether ketone) for proton exchange membranes." Journal of Membrane Science **229**(1–2): 95-106.

Ye, Y.-S., et al. (2012). "Water Soluble Polymers as Proton Exchange Membranes for Fuel Cells." Polymers **4**(2): 913-963.

Zhong, S., et al. (2008). "Modification of sulfonated poly(ether ether ketone) proton exchange membrane for reducing methanol crossover." Journal of Power Sources **180**(1): 23-28.

Fixed Plant Facility Siting – A Mega Project Lifecycle Framework

M.F. Haider
mech_haider@yahoo.com
mhaider@suncor.com

Prepared for Presentation at
American Institute of Chemical Engineers
2014 Spring Meeting
10th Global Congress on Process Safety
New Orleans, LA
March 30 – April 2, 2014

Fixed Plant Facility Siting – A Mega Project Lifecycle Framework

M.F. Haider
mhaider@suncor.com
mech_haider@yahoo.com

Keywords: Facility Siting, Project Management, Mega Project, Consequence Analysis (CA), Quantitative Risk Assessment (QRA), Adaptive, Change driven, Agile methodology

Abstract

Facility Siting based on tools such as Consequence Analysis (CA) and Quantitative Risk Assessment (QRA) is often times successfully applied for Brownfield operating unit projects, projects with relatively smaller footprints and for projects lasting relatively short time-periods. Challenge remains to apply and realize the full potential and value of Facility Siting for Mega-Projects. This is mostly due to project complexity and length – this paper will introduce a Management Framework to successfully apply CA and QRA as part of Facility Siting for Mega-Projects. The paper will also outline key strategies to ensure successful (i.e. value adding) implementation of facility siting, suggest a list of activities, associated sequence and pitfalls to avoid. Project Management team (e.g. Project Sponsor, Project Engineers, Managers) of both owner/operator and contractor/vendor may find this paper more useful as it provides a broad overview (i.e. when to initiate, how to plan, execute and close such assignment to maximize value) including real life examples of facility siting process over the project lifecycle.

1. Introduction

The driving force for process safety has been primarily based on catastrophic events [1]. With an increasing number of incidents, the process industry and governments started to take measures such as Facility Siting initiatives with hope to minimize loss of life and property. In the 1970s and 1980s, some of the world's most tragic industrial accidents took place. The Flixborough explosion in 1974 was reinforced by that of the Seveso tragedy in 1976 [2]. The Macondo blowout in the Gulf of Mexico on April 20, 2010, killed 11 employees and led to an uncontrolled oil spill lasting 87 days. This blowout was the most significant offshore incident in the US to-date. Finally, the Fukushima Daiichi nuclear plant incident in Japan in March 2011 drew the attention of the global process and power industries and encouraged them to incorporate natural disaster risks in a hazard analysis study [3]. As part of inherently safer design, many companies now-a-days apply CA and QRA as part of facility siting for their new projects. Due to the scarcity of information at the beginning, the studies in many cases start later in the project lifecycle for them to be effective and value adding. This paper will focus more on how to manage such initiatives for new Mega-Projects over the technical intricacies of the CA/QRA processes. The intention of the paper is to enhance the Body of Knowledge publicly available for management team in as much non technical languaga as possible to make their greenfiled Mega-Project initatives a success i.e. safer for people, environment and property.

So, what is different for industrial Mega-Projects?

- Industrial Megaprojects, as the name implies, are very large. To provide a simple and simply applied definition, we are defining a megaproject as any project with a total capital cost of more than $1 billion (U.S. dollars) as measured on January 1, 2003 [4].
- Most often engineering and construction activities run in parallel and not in sequence
- Multiple sub-projects under the 'Mega-Project umbrella' run in parallel
- Complex – design is almost 'never' complete in early stages – goes through many iterations
- Progressive elaboration is evident
- Scarcity of critical workforce with mega-project experience
- Scarcity of information – requires project team to make assumptions for CA/QRA
- Timeline: 2-5 years

2. Key Strategies

2.1 Facility Siting Charter

It is not only important or good to do but critical that the stakeholders need to identify, understand and agree on the purpose and value to conduct facility siting for their projects. The Author's personal experience over the years has been in order to maximize value of Facility Siting the project team has to start 'early' to develop project Charter and start with the end in mind. As can be seen from the following generalized Figure 1 that as the company starts to appraise opportunities, it is

Figure 1: Stages and Milestones to apply CA/QRA in a Mega-Project

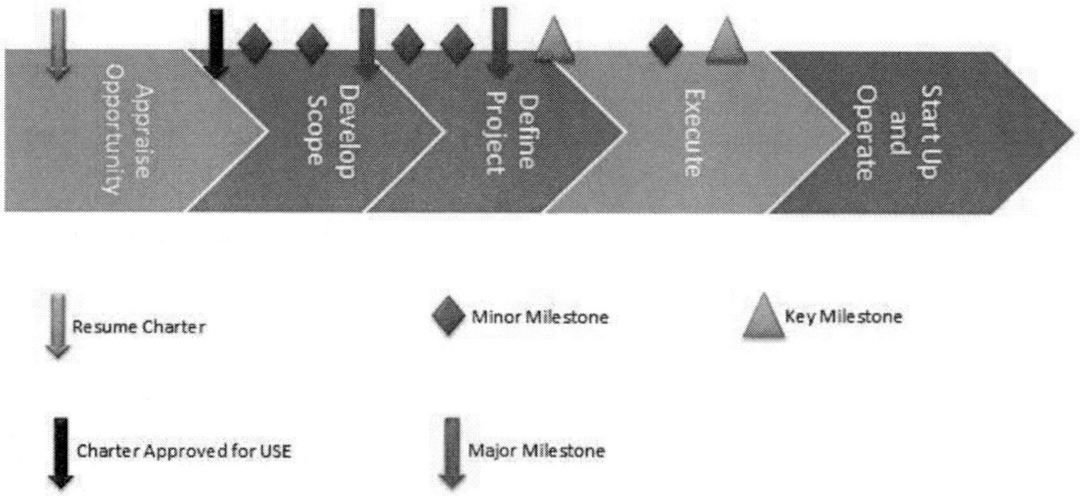

prudent to resume developing the Facility Siting Charter by bringing in the stakeholders together. Figure 1 shows a generalized project lifecycle for a typical Petro-Chemical Mega-Project. The

Figure also shows that the Key Deliverable is developed through a successive series of repeated cycles over the project life-cycle. The Milestones will be defined later in the paper. Part of starting with the end in mind is to clearly outlining ultimate goal/s, e.g. vision is to design a safer facility and utilize Mitigated QRA Report to formally document and communicate that facility is within the Risk Threshold or justification for accepting risk profile as is by Senior Management. The Vision in the Charter also provides opportunity to strike a balance between business goals and climbing the Risk Mountain – at each stage of increasing safety (decreasing risk). As part of striking a balance, the Project Engineer has to Identify options up-front if and/or when it becomes apparent that aggregate Residual Risk will not be within the Risk Threshold or acceptable so that team can put more effort in developing justification for either accepting or rejecting the risk profile by Management

Figure 2: Strike a balance between Business goal and Climbing the *expensive* Risk Mountain

2.2 Management Strategy

In order to successfully manage Facility Siting, CA and QRA for a Mega-Project, the stakeholders have to align on a management strategy. There are two primary management approaches for the purpose of managing such work:

1. Predictive or Plan driven
2. Adaptive or Change driven

The Author's past experience has been if the organization undertook very 'similar' projects in the near past (<5 years), critical experienced workforce is retained, Lessons Learned is readily available, and most importantly, the past project is still considered to be a success, it is advisable to use Plan driven approach. On the other hand, if the organization is introducing new technology, has limited past experience with similar project, involves relatively high number of stakeholders, capital cost > $1B, project life spans >2 years, it is recommended to utilize Change driven approach to maximize value. It is to be noted that it is not essential to meet all the suggested criteria mentioned above to pick one management strategy over the other.

2.2.1 Benefits of Adaptive or Change Driven Approach

In Adaptive/Change Driven Planning, project team intentionally repeats one or more project activities as the project team's understanding of the product increases ....Also known as ...agile methods .....intended to respond to high levels of change and ongoing stakeholder involvement [5]. The heart of Facility Siting involves making timely but informed decision. Some of the characteristics of this framework:

1. To add small incremental improvements that will deliver value to stakeholders
2. Milestones develop the product (e.g. Mitigated QRA Report) through a series of repeated cycles, while increments successively add to the functionality of the product
3. For each Milestone, a deliverable (read 'model') or set of deliverables will be completed. Future iterations may enhance those deliverables or create new ones. Each iteration incrementally builds the deliverables until the exit criteria for the phase are met, allowing the project team to incorporate feedback.

The biggest challenge with Mega-Project is scarcity of key information needed as inputs to develop scenarios and models. Some of the critical benefits, a project may gain by applying Change Driven approach:

1. The approach acts as Leading Indicator on Process Safety for Plant Layout - allows for multiple opportunities to make critical decisions on plant layout over the project life-cycle capitalizing on what is known at a given time. This lets the project team to use "what they know", to ensure "what they don't know" falls into the right place
2. Opportunity to make decisions over the project lifecycle not only works to save life but also helps with striking a balance between business needs and climbing the risk mountain
3. Aides design-development with progressive elaboration while producing final CA/QRA deliverable for the project

2.3 Execution Strategy

The main idea of Change Driven process starts with decomposing the known preliminary layout and process information into a set of requirements, work to be performed and their priority sequence, sometimes referred to as a *Product Backlog*.

It is advisable that the owner/operator forms a multi-use agreement (contract) with the chosen vendor. This way the plant areas under the Mega-Project umbrella can get their specific work done while the CA/QRA Project Team provides overall oversight and guidance.

At the beginning of an iteration, the team will work to determine how many of the highest priority items on the backlog list can be delivered within the next iteration. The time-length of each iteration should be less than 4 weeks. As shown in Figure 1, during "Develop Scope" and "Define Project", the length of each iteration could be even less than 2 weeks. This iterations (otherwise known as "Minor Milestone" as shown on Figure 1) will develop scenarios, assumptions and preliminary CA/QRA models. Assumptions are required due to lack of available quality information in the early stages of a project.

At the end of each iteration, the Minor Milestone/deliverable should be ready for review by the customers/stakeholders. This does not mean that the customer is required to accept delivery, just that the product should not include unfinished and/or incomplete features.

The sponsor and customer representatives should be continuously engaged with the other project stakeholders to provide feedback to Contractor on deliverables as they are created and to ensure that the *Product Backlog* reflects Sponsor/stakeholders' current needs. Below, you will find two examples using two different management strategies. For confidentiality purpose, the specific name/location is kept generic and few items were modified to suit.

2.3.1 Project X Facility Siting Using Change Driven Approach

A CA/QRA team was given the assignment to support design development during "Develop Scope" phase (Figure 1). The Sponsors already selected the site for the project. But the fixed plants were yet to be located on the site. The Vision for the initiative was for the QRA result to come under the organization Risk Threshold. Multiple EPC's were given to engineer different plant areas. The CA/QRA team selected a vendor to support them with defining the *Product Backlog* based on the preliminary Process Design Basis, intended product the facility is expected to sell, the feedstock necessary etc. The CA/QRA team had prioritized the Product Backlog based on hazard level of feedstock, preliminary PHA schedule and project schedule. This also involved obtaining feedback from different Process Engineering and Operations teams to collect necessary inputs to put together the assumptions and scenarios to be modeled within the 1st iteration.

Out of 25+ scenarios, only 7 perceived to be the most hazardous were modeled for CA that gave ample input (e.g. explosion, toxic and fire contours) for the CA/QRA team to work with the stakeholders to recommend location for specific process units on the given site. Note that, CA/QRA team integrated the potential consequence to understand local incident and global consequence events. At this time QRA was not performed. This was used as input for minor modification of preliminary plant layout. It is to be noted that the site plan was still preliminary after the first iteration but assisted different EPC accountable for their specific area to recognize where to locate the specific plants i.e. finalized key plans for the overall site plan.

With further design-development, more refined information became available from each EPC. This new information led to 130+ scenarios and was used for further CA/QRA iterations that created either new model and/or embellished the past ones. Each EPC was accountable for facility siting within their boundary while the CA/QRA team provided technical oversight and integration support for the overall project. The Sponsor/stakeholders were engaged on a regular basis to ensure that the Product Backlog reflects future modeling requirements to properly site facilities e.g. Fire Water Pump house, Tank Farms, site Medical Centre etc. Early engagement of CA/QRA team and higher management willingness to incorporate their feedback with the progressive elaboration of the project led to QRA result meeting the organization risk threshold and ultimately safer facility. Changes suggested based on CA/QRA work in earlier stage (i.e. Develop Scope, Define Project) were easier to incorporate into the design.

2.3.2 Project Y Facility Siting Using Plan Driven Approach

The sponsors for project Y selected the site for the project and decided to hold on to not perform CA/QRA work with the progressive elaboration of the project activities. This was mostly decided based on the fact that not all information was available to properly conduct the 'studies'. Fundamentally, the sponsors looked CA/QRA as deliverables requiring a more 'complete' plan. Once the Develop Scope (as referred in Figure 1) is almost complete, the sponsors decided to engage CA/QRA team and a vendor. By this time, the project had significantly high number of stakeholders. Many of the stakeholders were not very familiar with CA/QRA and the value it could bring to them. There was not much time to schedule and train the stakeholders as well to bring them up to speed. As a result, the stakeholders were unable to align on a Vision. The Vendor was asked to provide explosion, toxic and fire contour maps for specific scenarios and locations that were almost 'fixed' on the site plan at that point in time. The resultant contour maps showed significant consequence on nearby facility with over 700+ people. It was also noted that the plant area consuming the most hazardous substance would impact the nearby facilities significantly from consequence perspective. It should be noted that the result was only based on a selected few scenarios. By the time the project reached the "Execution" phase, more scenarios were added as part of the scope. But it was already too late to incorporate most of the recommended changes resulting from CA/QRA work. The project went through several area by area risk reduction workshops only to accept (by the company executives per organization risk matrix) the aggregate residual risk level which was also found to be not within the organization risk threshold.

2.3.3 Importance of Starting Early

Figure 3: Impact of variable based on project time [5]

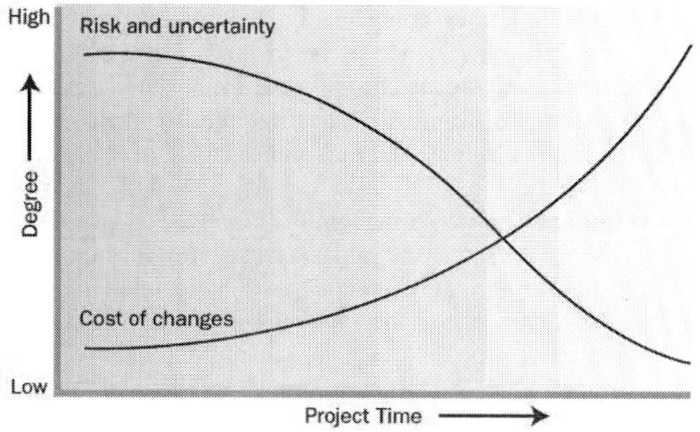

The above two examples highlight the importance of starting CA/QRA work early enough to realize and maximize the value CA/QRA brings to a Mega-Project. The high-level view provides common frame of reference for comparing projects for CA/QRA initiative, even if the projects were dissimilar. Change driven methodology allows for applying the known information at early stage in a way that supports laying out plant facilities to reduce risk and thus maximize potential value. While these characteristics remain present to some extent in almost all project life cycles, they are not always present to the same degree [5]. Change Driven framework, in particular, keeps stakeholder influences higher and the costs of changes lower especially in the early stages of the life cycle than in predictive/plan life cycles. The Figure 3 highlights a well-known fact of project management that even though risk and uncertainty remains high in the early stages of a project, cost of incorporating changes remains much lower. This translates into a key opportunity for project stakeholders to align on laying out the key areas on a site plan intelligently to reduce risk and maximize value by applying tools such as CA and QRA.

2.4 Key Decision Points

Facility Siting, from a management point of view, is all about bringing the stakeholders closer and find the 'right' spot for the facilities. Finding the 'right' spots requires making timely decisions. The challenge has always been to make timely decisions as the stakeholders more often find themselves forced to make those decisions based on information that are frequently of poor quality and at times uncertain, especially in the early stages of a project. At times, due to the scarcity of information, stakeholders rely on their past experience to make those decisions on plant layout. As a result, depending on the individual experience, two different teams for a project area may come up with two very different plant layout. CA/QRA work, if started early enough, will also depend on input that most often requires embellishment through assumptions to create needed scenarios and models. Even with this limitation, CA/QRA provides a key 'input' in allocating real estate for key facilities on a plant site at early stage of a project lifecycle. With each iteration and deliverable 'Change driven' methodology allows for multiple opportunities to make critical decisions on plant layout over the project life-cycle. It allows the stakeholders to decide when to 'sharpen the pencil' (e.g. PHAST to Computational Fluid Dynamics model), relocate facility or accept the consequence and/or risk, 'pull the plug' on further modeling/refinement and start developing justification, if required, to recommend management to accept/reject the aggregate risk profile.

2.5 Alignment with Process Hazard Analysis (PHA) Planning

CA/QRA results are a useful input to PHA (e.g. HAZOP, LOPA) activities. This is particularly true as projects move forward after establishing the plant layout on a high level and starts to conduct "Define Project" (Figure 1). PHA activities may also result in recommendations (i.e. new scenarios) that goes into the *Product Backlog* and requires project team to develop new models or refine old ones. Mega-projects have multiple sub-projects running in parallel most of the time, having an Integrated Schedule for all the reviews is a good strategy so that the right stakeholders are booked for the sessions. The integrated PHA schedule also supports prioritization of *Product Backlog* for CA/QRA activities over the lifecycle of the project.

2.6 Integration Strategy

It is critical for a Mega-Project involving multiple plant areas/silos to understand the BIG picture so that site risk can be dealt with on an aggregate basis. There is strong possibility of one area to impact other plants in the vicinity. It is recommended to review integration aspect and aggregated results on a regular basis. The following Figure 4 shows that the origin of an incident may reside in one particular plant area, while it may impact other areas in the vicinity.

Figure 4: Consequence of an incident in one area impacting multiple areas

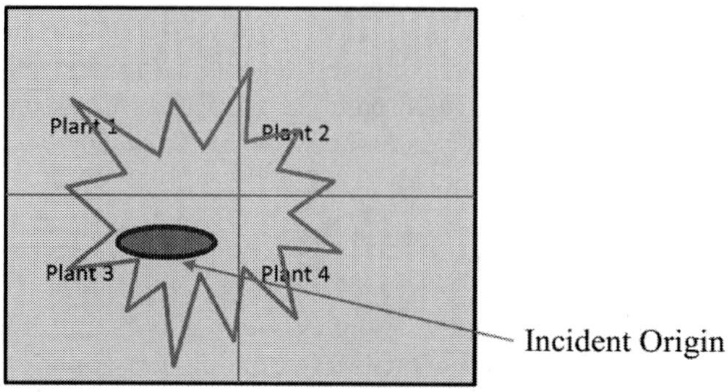

Recommended Strategy:

- Setup Strong Center - Task Force Single Point of Contact (SPOC)
- Collaborate – SPOC will promote Collaboration among Plants/Areas

3. List of Activities (suggested)

3.1 Overall Sequence

The CA/QRA work typically should start with the approved Charter as shown on Figure 1. A project Task Force i.e. SPOC is formed comprising of Project Engineer, Operations rep, Process rep and CA/QRA Technical Expert. The SPOC is ultimately accountable to produce the CA/QRA deliverable studies but also acts to support the design development and site plan finalization efforts in the early stages of the project.

The team collects 'input' such as Preliminary site Layout and design information, works with area process and operations team to identify scenarios, develop assumptions for known-unknowns and unknown-unknowns, develop modeling 'Product backlog', prioritize backlog based on PHA schedule and produce modeling output (i.e. deliverable) for stakeholder use. The stakeholder usually has the option of accepting or rejecting consequence. If the result of an iteration is acceptable within area causing the incident – communicate to outside stakeholders to ensure there is no cross boundary impact. If the result of the iteration is NOT acceptable to outside of the area – use the result as input to PHA for the areas to mitigate. Collaboration recommended between origin area and impacted area/s to identify appropriate mitigations. There is also a possibility that the consequence resulting from the iteration is not acceptable to the area causing the incident, in that case the area will initiate a PHA to mitigate the consequence. As part of developing mitigations, consider the following:

- Develop QRA if appropriate
- Perform Building Assessments for Critical Buildings
- Apply Cost-Benefit analysis to optimize risks/opportunities

- Determine Safe Haven locations and their characteristics

If Mitigation is not possible from Business perspective – perform As Low As Reasonably Practicable (ALARP) justification for Senior Management to formally accept the posed risk
- To apply mitigations on a site wide basis, review Milestone deliverables by taking on a BIG picture perspective
- Future iterations may enhance those deliverables or create new ones as new information is available. Each iteration incrementally builds the deliverables to produce the Mitigated CA/QRA report to support Facility Siting for the stage

The overall effort is expected to produce a maximum of two single reports for the overall site:

1. Mitigated version of QRA Report
2. and Consolidated CA Report

It is recommended to develop as-built, pre-start up CA/QRA reports prior to start-up. Table 1 and 2 outlines typical activities performed by SPOC and plant areas respectively once key process areas are established/finalized on Site Plan. Please note that the activities are project dependent and may warrant modifications to suit specific situation.

Table 1. SPOC Sequence of Activities to conduct Facility Siting

Initiative	Key Steps
Setup Facility Siting Team	Sponsor to support to setup a team (i.e. Project Engineer, Technical Expert, Operation, Process representative) to act as SPOC
Integration	Provide functional (technical and management) guidance to Areas
	Assess cross boundary impact/opportunity
	Communicate incremental outputs after each Area PHA/Risk Reduction
IFA versions (prior to "Define Project" stage)	Consolidated CA Report
	Mitigated and Integrated QRA Report
IFD CA/QRA (prior to "Execution" stage)	Re-validate IFA CA/QRA post PHA and vendor PHA
As-built versions	Post-oil version

3.2 Typical Area Activities

Table 2. Area (Typical) Sequence of Activities to conduct Facility Siting

Initiative	Key Steps
Plant 1 (Typical)	Perform Modeling and Calculations
	Conduct Risk Reduction Workshop (if appropriate) – Apply mitigations
	Incremental output for use based on reduced risk level
	Assess Buildings/facilities
	Identify Safe Haven/shelter Locations (if required)
	Estimate Area Project Scope impact
	Perform area scope updates/changes e.g. passive fire protection requirement
	Carry out QRA (if appropriate)
	Update Product Backlog
	Develop/update contract to support area specific activities

4. Pitfalls to Avoid

The following observations were noted by the Author during his time to manage Facility Siting scope:

1. Facility Siting does not replace good and common sense engineering practice e.g. locating a plant that uses high volume of hazardous material will significantly increase risks to other plants/areas (e.g. camp) within the vicinity
2. Evaluate effectiveness of 'Cost Reduction' philosophies as opposed to added risk e.g. different modularization techniques may decrease the plant footprint by reducing pipe lengths but makes the plant more compact and likely to increase severity of explosion consequence
3. Avoid developing models with hope to realize risk reduction
4. Sharpening the pencil for models most likely may not reduce risk
5. Avoid jumping into CFD or other similar methods without evaluating cost-benefit for undertaking such effort
6. Utilization of QRA results is much more controversial than the methodology. Watson [6] has suggested that QRA should be considered as an argument, rather than a declaration of truth. In this view – focus on accuracy over precision
7. CA/QRA Task Force should be part of Management of Change (MOC) process
8. Evaluate carefully before locating facilities e.g. tank farms close to main entrance/egress as part of Facility Siting

9. Keep the focus on VISION for the overall effort to avoid scope creep
10. Note the difference between Facility Siting as a Deliverable and tool for design-development activities

5. Conclusion

This paper presented Adaptive/Change driven framework to manage initiatives like facility siting/CA/QRA as part of Mega-Projects. Change driven framework does not mean "lack of planning or no planning". As demonstrated in the paper, this methodology starts with a vision and then develops the ultimate product both iteratively and incrementally when it is possible to define small incremental improvements that will deliver value to stakeholders. This allows making key decisions on laying out the facilities in early stages of a project rather than waiting for design to develop to a significant extent. Often times, project team wait too long for design to mature enough before applying CA/QRA to validate consequence impact and/or risk level. Most often, especially for Mega Projects, this leads to delays in executing the CA/QRA work properly at the right time to influence site layout. This delay in execution leads to scenarios when scope changes cannot be incorporated anymore. This translates into value erosion as project was unable to reap the full benefit of Facility Siting. The management framework recommended in this paper i.e. Change driven/adaptive framework allows for the stakeholders to start early in the project lifecycle to apply CA/QRA in laying out the plant facilities. The key is to start EARLY but START with the END in mind. Alignment among stakeholder on a VISION up-front is also important as this is expected to save life while striking a balance between business needs and climbing the Risk Mountain. Vision also allows avoiding scope creep while applying Adaptive/Change driven approach. It is highly recommended that stakeholders focus more on accuracy over precision. In the early stages of project lifecycle, quality information is scarce. Change driven/Adaptive framework gives you as the project stakeholder the opportunity to use "what you know" intelligently to ensure "what you don't know" falls into the right place.

6. References

[1] M. Mannan, A. Chowdhury and O. Reyes-Valdes, "A portrait of process safety: From its start to present day," College Station, 2012.

[2] M. Mannan, Lees' Loss Prevention in the Process Industries, Volumes 1–3 (3rd Edition), , 2005, Elsevier, 2005.

[3] K. McAndrews, "Consequences of Macondo: A Summary of Recently Proposed and Enacted Changes to US Offshore Drilling Safety and Environmental Regulation," in *Society of Petroleum Engineers, Americas E&P Health, Safety, Security and Environmental Conference*, 2012.

[4] E. W. Merrow, Industrial Megaprojects Concepts, Strategies and Practices for Success, New Jersey.: John Wiley & Sons, Inc., 2011.

[5] A Guide to the Project Management Body of Knowledge (PMBOK® Guide) – Fifth Edition, Newtown Square: Project Management Institute, Inc., 2013.

[6] S. Watson, "The Meaning of Probability in Probabilistic Risk Assessment," *Reliability Engineering and System Safety,* vol. 45, pp. 261-269, 1994.

348480 Global Energy and Transportation and Mobil Oil

Abstract:

Biofuels, Computational Methods, Engineering Economics, Polymers, Thermodynamics. Polymers-plastic, Thermodynamics- oil, Engineering Economics – the money it takes to finish the method of Global Engineering Research, Computational Methods – numerical input of analytical data, Biofuels – the transport fluids that utilized in more than one method from industrial usage to everyday living which enters the atmosphere. Exxon has completed a $600 million dollar deal for production of algae biofuel, amongst Mobil Oil has performed major research Global Energy and the effects of oil. Reading a most recent article on Mobil website.

348748 Characterization of Iron Phthalocyanine As the Active Material for Lithium Batteries

Abstract:

Results on the determination of the lithium-ion diffusion coefficient in the iron phthalocyanine (FePc) active material as a function of intercalation level, using a modified electrochemical impedance spectroscopy technique, will be presented. Also, developed thermodynamic functions such as the Gibbs free energy, entropy, and enthalpy changes for the lithium insertion reaction with FePc, derived from cell open-circuit voltage measurements as a function of intercalation level, are presented. Also presented are estimates of the maximum amount of lithium that can be intercalated into the FePc based on constant volume and temperature conditions.

349086 Design of a Free-Fall Reactor for Fast Pyrolysis of Waste Plastics

Abstract:

Thermal pyrolysis has earned significant reputation in the recovery of valuable chemicals, especially petrochemical based fuels from various materials, including biomass and plastics. Of many available technologies, *fast* pyrolysis was preferred in his study owing to its minimal heat and mass transfer limitations and better control on process conditions. The focus of this study is towards efficient conversion of waste plastics into gaseous products consisting primarily of alkanes, alkenes and aromatics. Lab scale experiments were performed using a micro reactor to determine favorable conditions, including reaction temperature and particle size that would result in maximum yield of ethylene. Using these results, a simple industrial size radiatively heated free-fall reactor was designed for the purpose of pyrolyzing waste plastics. The goal is to design a *fast* pyrolysis reactor that improves upon inefficiencies in traditional reactors while resulting in high yield of ethylene.

349732 Biosorptive Dehydration of Ethanol/Water Azeotropes Using Compound Starch-Based Adsorbent

Abstract:

Abstract: Recently, the interest in the production of bioethanol has greatly increased as a result of the increasing fossil fuel prices and associated environmental issues. Bioethanol is made from either grains or biomass and the broth contains 6 to 12 percent ethanol. In addition, water removal represents a serious problem as ethanol form azeotrope with water at atmospheric pressure. Distillation to water-free ethanol consumes 50 to 80 percent of energy used in the fermentation ethanol manufacturing process. Various techniques have been developed to break the azeotropic point, such as salting-out method, azeotropic distillation, extractive distillation, reactive distillation, adsorptive distillation and pervaporation. Among the methods, the adsorptive option is the most attractive, especially the biosorption, which has distinct advantages over conventional methods as it is environmentally friendly, highly selective, easily available, easy to operate, cost effective and reusable in repeated cycles in the treatment of aqueous organics.

In this work, a specially formulated compound starch-based adsorbent (CSA), which consists of corn, sweet potato and foaming agent, was developed. As an exploratory study, this research was focused on the investigation of CSA for biosorptive dehydration of aqueous ethanol and optimization of conditions for its maximum adsorption under various conditions of vapor feed flow rate, bed temperature and bed height through an orthogonal design method. Furthermore, to investigate the adsorption mechanism, dynamic saturated adsorbance of CSA was studied and characterization of the adsorbent before and after adsorption, as well as after regeneration, was performed.

In the first step, the net retention time and separation factor of ethanol and water were obtained using inverse gas chromatography (IGC), the results indicated that the net retention time of water on CSA is significantly longer than that of ethanol which attested the feasibility of this adsorptive separation purpose and low temperature could be propitious to the process which was consistent with the traditional theory of adsorption.

Then, through a batch adsorption experiment, the optimum adsorption condition was determined under vapor feed flow rate, bed temperature and bed height by orthogonal design method, which indicated that the optimization condition for ethanol was at vapor feed flow rate of 1.1g/min, bed temperature of 83   and bed height of 25cm. The packing density was 445kg/m^3. The residence time is determined by vapor feed flow rate and bed height. The residence time is more influenced than bed temperature. The longer the residence time, the more the output is. But it cannot be ignored that the adsorption selectivity may decreases. Low bed temperature is benefited to the adsorption process which agrees with the result of IGC. Five cycles experiments of adsorption-regeneration were carried out at the optimum adsorption and regeneration condition, the results showed that there is practically no change in the regeneration efficiency of CSA after the cyclic process adsorption–regeneration. It is due to the properly chosen regeneration conditions assuring complete desorption of the water adsorbed without changes in the CSA structure. This unchanged adsorption capacity after a number of cycles demonstrates reusability of the adsorbent.

To study the adsorption mechanism, experiment was conducted to study the dynamic saturated adsorbance of CSA. The water adsorption capacity of CSA has been found to be similar to some molecular sieves, which is about 0.15g/g. Besides, field emission scanning electron microscopy (FESEM) and mercury porosimetry (MP) were applied to investigate the change of surface morphology and microstructure of CSA before and after the adsorption, as well as after regeneration, as shown in Fig.1 and Fig.2. The FESEM images showed that compared with the particles before the adsorption, they became plump and the void spaces became smaller after adsorption, which indicated that the CSA was saturated with water. However, the CSA particles after regeneration became wizened like the nature particles, which demonstrated that almost all the water had been desorbed from CSA. The pore size distribution (PSD) indicated that compared with the PSD curve before adsorption, the peak at 6000nm disappeared and the peak at 17000nm became remarkably smaller after adsorption, which manifested that CSA is mainly composed of macropores which play a major role in the adsorption. However, the peaks at 6000 and 17000nm reappeared after regeneration, and the curve looked exactly the same with that before adsorption, which further demonstrated that CSA after regeneration can be totally regenerated and recovered to the original structure.

These results indicated that the starch-based adsorbent can be considered as an effective and low-cost bio-sorbent to dehydrate alcohol.

349781 Purification of 2-Amino-1-Phenylethanol Enantiomers By a New Technique Combining Distillation and Crystallization

Abstract:

Stripping crystallization (SC) is a new separation technique which combines distillation and crystallization. Basically, SC is operated at a triple-point condition, in which the liquid is simultaneously vaporized and crystallized due to the three-phase equilibrium. In essence, the process is continued until the liquid phase is completely eliminated and only the pure solid crystals remain in the feed. Thus, crystal washing is not required since only pure solid crystals remain in the feed and no impurities are adhered on the crystal surfaces at the end of SC. In the previous research, SC has been successfully applied to separate p-xylene from the mixed xylenes by Shiau et al. (2005, 2008).

Phenylethanolamine, also known as 2-amino-1-phenylethanol (2-APE), is an important basic skeleton in many adrenergic drugs and is often used as ingredients of cold medicines and anorectics. As the biological activity and pharmacokinetics of the enantiomers of a chiral drug can be different, the objective of this research is to apply SC to purify R-2-APE from the chiral 2-APE mixture. The experiments show that no crystallization occurs during SC without seeding due to subcooling of the liquid enantiomers. On the other hand, seeding can eliminate subcooling during SC, leading to the R-2-APE crystals gradually formed preferentially on the seeds. Thus, seeding with magnetic stirring can be successfully employed to purify R-2-APE from the 2-APE enantiomers.

1. Shiau, L. D., Wen, C. C. and Lin, B. S. Separation and purification of p-xylene from the mixture of m-xylene and p-xylene by distillative freezing, *Ind. Eng. Chem. Res*, **44(7)**, 2258-2265, 2005.

2. Shiau, L. D., Wen, C. C. and Lin, B. S. Separation of p-xylene from the multicomponent xylene system by stripping crystallization, AIChE Journal, 54(1), 337-342, 2008.

350841 Fabrication of Low Cost Insulating Material from Kaolin Clay for Construction Purposes

Abstract:

A new low cost, thermal insulator, construction mineral polymeric material was synthesized from local Saudi Arabia resources using Inorganic Mineral Polymerization technique (MiP). In this technique, the mineral polymer is produced from the following components: clay mineral rich in koalinite, sand, sodium hydroxide solution, and water. Inorganic polymerization experiments are conducted by mixing known amounts of clay minerals and sand in a variable speed stainless-steel mixer for some period of times until a homogenous solid mixture is obtained. Sodium hydroxide solution is added to the mixture and the polymerization reaction takes place at different mixing speeds and times. The produced mixture is divided into a number of samples of known weights. Each sample is moulded in a stainless-steel cylinder at a medium pressure. Then the moulded samples are cured in an oven for 24 hours. The curing temperature was 80°C. The mineral polymerization conditions such as the ratio of clay to sand filler, particle size of the clay, particle size of the sand, concentration and amount of sodium hydroxide, speed and time of mixing were optimized to obtain stable and hard mineral polymer. Compressive strength for different specimens of fabricated mineral polymers was tested under dry and wet conditions. Values of compressive strength with values ranging from 11.5 MPa to 27.9 MPa were obtained.

351141 Thermodynamic Study of Binary PAH (Anthracene + Phenanthrene) Solid Mixtures

Abstract:

Polycyclic aromatic hydrocarbons (PAHs) are common components of many materials, such as petroleum and various types of tars. They are generally present in mixtures, occurring both naturally and as byproducts of fuel processing operations. It is important to understand the thermodynamic properties of such mixtures in order to better understand and predict their behavior (i.e., fate and transport) in the environment and in industrial operations. The Knudsen effusion technique was combined with gas chromatography and differential scanning calorimetry to measure the vapor pressure, composition, and phase behavior of binary anthracene and phenanthrene mixtures at near ambient temperatures. Mixtures of anthracene and phenanthrene exhibited non-ideal mixture behavior: In mixtures with less than 80% phenanthrene, this more volatile component preferentially sublimed and the vapor pressures decreased as the mixtures became enriched in anthracene. The vapor pressure data do not follow Raoult's law (that is, ideal mixture) predictions. Mixtures with high phenanthrene content (> 80%) showed azeotrope like behavior, in that anthracene and phenanthrene sublimed at nearly constant vapor pressure and molecular ratio. These results confirm that the behavior of few-component PAH mixtures does not follow ideal mixture models. Hence, as tars and petroleum (and potentially other organic compound-containing mixtures) weather to few component mixtures over time, complicated, non-ideal phase behavior might be observed.

351160 How Confident Are You That Your Major Accident Risks Are Under Control?

Abstract:

Major accidents in the process industries including those at Texas City , Buncefield UK and in the Gulf of Mexico have highlighted the critical role played by senior managers and executives. Effective leadership is essential if companies are to develop a positive safety culture that remains constantly vigilant towards the risk of a major process safety incident. In the UK the HSE has made leadership a key priority for the high hazard industries and there is a cross-industry initiative to develop and deliver training focussed towards board members responsible for process safety performance.

This paper explores the leadership failings that contributed to recent major accidents, and the cost to both the company reputation and the individuals involved. Key principles are discussed that can help improve leadership in the major hazard industries. These include; attitudes and decisions of senior manager affecting the investment strategy and the safety culture of the organisation, reinforcing the importance of safety by personal example, thorough understanding of major accident hazards and key risk control systems, investigating process safety incident and near misses to find the underlying causes, developing world class safety management systems, and identifying weaknesses in these systems using targeted performance indicators.

The author bases this paper on the development of a training standard for 'Process Safety Leadership' involving a cross-industry 'expert panel'. Training of senior executives in compliance with this standard has been carried out from late 2011 and feedback from these courses and the key requirements for effective leadership will be discussed.

351236 Sil Determination: Shortcomings with the Use of LOPA

Abstract:

Proposed Topic Area: Managing Risks

For a number of years, industry has been trying to apply the principles outlined in IEC 61508. In the process sector, this has resulted in a focus on the requirements of IEC 61511 (2003) and ANSI/ISA 84.00.01 (2004). These are risk based standards. They involve the identification of potential hazardous events, the causes that could lead to each event and the necessary risk reduction for each event. Where the risk reduction involves an instrumented function (e.g. a trip or an interlock), the required performance of the instrumented function can described in terms of the average probability of failure on demand. In the standards, the performance for such instrumented functions is assigned to one of four Safety Integrity Levels. The assessment of what performance level is needed in respect to a specific hazardous event is referred to as SIL Determination. In the field of SIL Determination, IEC 61511 provides examples of a number of different approaches – these are shown in IEC 61511 Part 3. These include Event Trees, Risk Graphs, Safety Matrices, Fault Tree Analysis and Layer of Protection Analysis (LOPA).

In the first few years, following the publication of IEC 61508 Edition 1 in 2000, industry generally embraced the use of Risk Graphs as the method of choice. However, over the subsequent few years, many companies using this method found that it was (a) giving higher safety integrity level targets than really needed and (b) was less flexible than other methods and consequently more difficult to use in many real situations. This has resulted in a general move to other methods and in particular to the use of Layer of Protection Analysis. Indeed, LOPA may now be the predominant "Method of Choice" for SIL Determination. As a method, LOPA gives the impression of simplicity and appears quite straightforward. It almost invites anyone with a minimum of understanding of the method to be a practitioner.

However, there are a number of shortcomings in the use of the LOPA method that have become evident through its use across the globe.

This paper will review the application of LOPA and provide insights into what may be seen as the proper use of the method and its core principles. It will also highlight various shortcomings in use that have been found during expert reviews carried out of a number of existing studies. The paper will be of benefit both to those who are new to the LOPA method, but also of value to those who have been using the method for some time.

AUTHOR INDEX